Process Control

Process Control
Pressure, Flow, and Level

Student Manual
585112

Order no.: 585112
First Edition
Revision level: 07/2015

By the staff of Festo Didactic

© Festo Didactic Ltée/Ltd, Quebec, Canada 2009
Internet: www.festo-didactic.com
e-mail: did@de.festo.com

Printed in Canada
All rights reserved
ISBN 978-2-89640-371-4 (Printed version)
ISBN 978-2-89640-917-4 (CD-ROM)
Legal Deposit – Bibliothèque et Archives nationales du Québec, 2009
Legal Deposit – Library and Archives Canada, 2009

The purchaser shall receive a single right of use which is non-exclusive, non-time-limited and limited geographically to use at the purchaser's site/location as follows.

The purchaser shall be entitled to use the work to train his/her staff at the purchaser's site/location and shall also be entitled to use parts of the copyright material as the basis for the production of his/her own training documentation for the training of his/her staff at the purchaser's site/location with acknowledgement of source and to make copies for this purpose. In the case of schools/technical colleges, training centers, and universities, the right of use shall also include use by school and college students and trainees at the purchaser's site/location for teaching purposes.

The right of use shall in all cases exclude the right to publish the copyright material or to make this available for use on intranet, Internet and LMS platforms and databases such as Moodle, which allow access by a wide variety of users, including those outside of the purchaser's site/location.

Entitlement to other rights relating to reproductions, copies, adaptations, translations, microfilming and transfer to and storage and processing in electronic systems, no matter whether in whole or in part, shall require the prior consent of Festo Didactic GmbH & Co. KG.

Information in this document is subject to change without notice and does not represent a commitment on the part of Festo Didactic. The Festo materials described in this document are furnished under a license agreement or a nondisclosure agreement.

Festo Didactic recognizes product names as trademarks or registered trademarks of their respective holders.

All other trademarks are the property of their respective owners. Other trademarks and trade names may be used in this document to refer to either the entity claiming the marks and names or their products. Festo Didactic disclaims any proprietary interest in trademarks and trade names other than its own.

Safety and Common Symbols

The following safety and common symbols may be used in this manual and on the Lab-Volt equipment:

Symbol	Description
⚠ DANGER	**DANGER** indicates a hazard with a high level of risk which, if not avoided, will result in death or serious injury.
⚠ WARNING	**WARNING** indicates a hazard with a medium level of risk which, if not avoided, could result in death or serious injury.
⚠ CAUTION	**CAUTION** indicates a hazard with a low level of risk which, if not avoided, could result in minor or moderate injury.
CAUTION	**CAUTION** used without the *Caution, risk of danger* sign ⚠, indicates a hazard with a potentially hazardous situation which, if not avoided, may result in property damage.
⚡	Caution, risk of electric shock
♨	Caution, hot surface
⚠	Caution, risk of danger
🏋	Caution, lifting hazard
✋	Caution, hand entanglement hazard
⎓	Direct current
∼	Alternating current
⎓∼	Both direct and alternating current
3∼	Three-phase alternating current
⏚	Earth (ground) terminal

Safety and Common Symbols

Symbol	Description
⏚	Protective conductor terminal
⏚	Frame or chassis terminal
⏚	Equipotentiality
│	On (supply)
○	Off (supply)
▣	Equipment protected throughout by double insulation or reinforced insulation
⊓	In position of a bi-stable push control
⊓	Out position of a bi-stable push control

Preface

Automated process control offers so many advantages over manual control that the majority of today's industrial processes use it to some extent. Breweries, wastewater treatment plants, mining facilities, and the automotive industry are just a few industries that benefit from automated process control systems.

Maintaining process variables such as pressure, flow, level, temperature, and pH within a desired operating range is of the utmost importance when manufacturing products with a predictable composition and quality.

The Instrumentation and Process Control Training System, series 353X, is a state-of-the-art system that faithfully reproduces an industrial environment. Throughout this course, students develop skills in the installation and operation of equipment used in the process control field. The use of modern, industrial-grade equipment is instrumental in teaching theoretical and hands-on knowledge required to work in the process control industry.

The modularity of the system allows the instructor to select the equipment required to meet the objectives of a specific course. Two mobile workstations, on which all of the equipment is installed, form the basis of the system. Several optional components used in pressure, flow, level, temperature, and pH control loops are available, as well as various valves, calibration equipment, and software. These add-ons can replace basic components having the same functionality, depending on the context. During control exercises, a variety of controllers can be used interchangeably depending on the instructor's preference.

We hope that your learning experience with the Instrumentation and Process Control Training System will be the first step toward a successful career in the process control industry.

Standard Learning Path

Familiarization → Measurement → Process Control → Advanced Process Control

Specific equipment User Guide or Manual
(transmitter, valve, calibration equipment, controller, or software)

We invite readers of this manual to send us their tips, feedback, and suggestions for improving the book.

Please send these to did@de.festo.com.

The authors and Festo Didactic look forward to your comments.

Preface

Manuals of the 353X Series

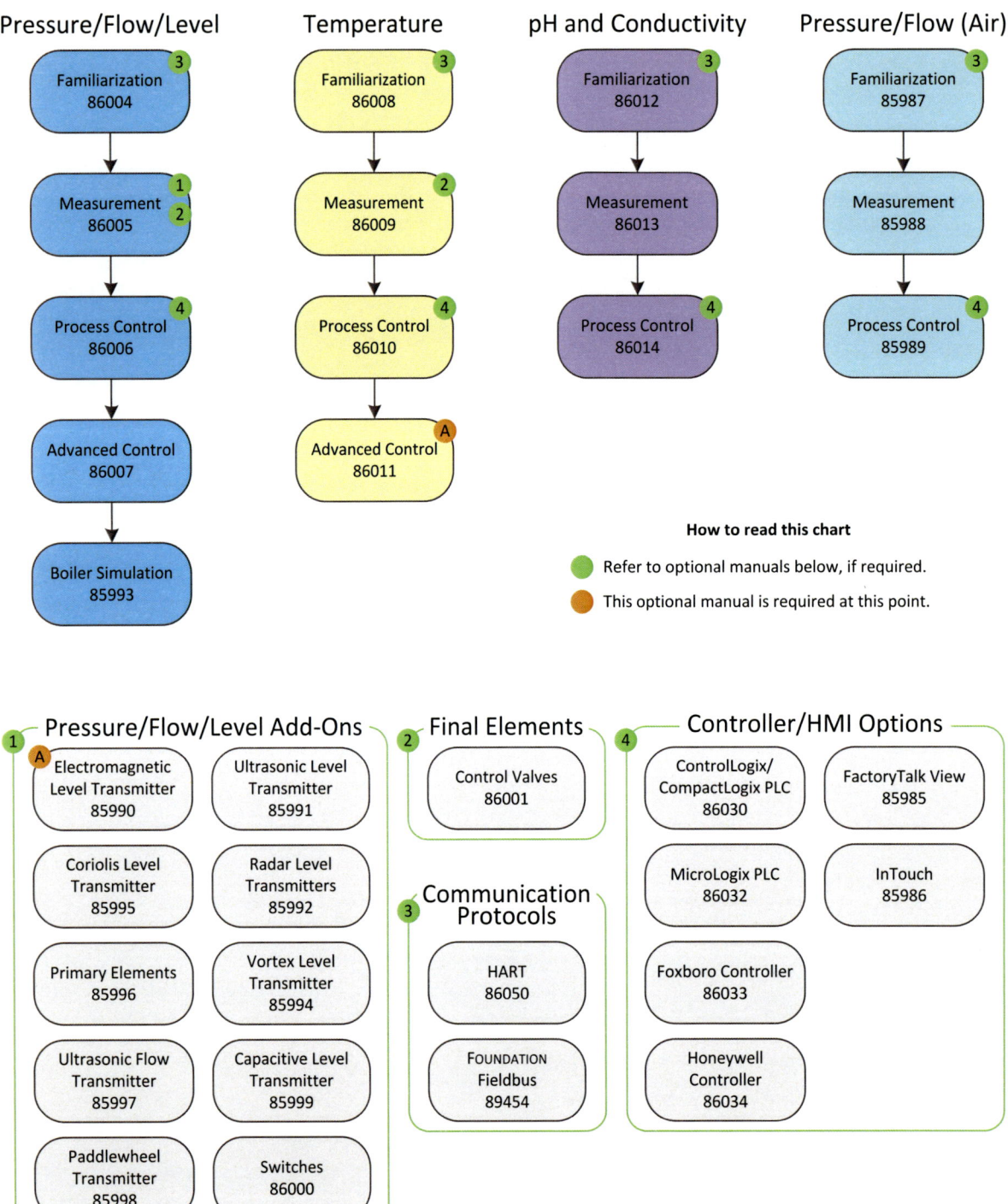

Table of Contents

Unit 1 **Process Characteristics** 1

Ex. 1-1 **Determining the Dynamic Characteristics of a Process** 15

Unit 2 **Feedback Control** 31

Ex. 2-1 **Tuning and Control of a Pressure Loop** 53

Ex. 2-2 **Tuning and Control of a Flow Loop** 65

Ex. 2-3 **Tuning and Control of a Level Loop** 77

Ex. 2-4 **Cascade Control of a Level/Flow Process** 87

Unit 3 **Troubleshooting a Process Control System** 103

Ex. 3-1 **Guided Process Control Troubleshooting** 109

Ex. 3-2 **Non-Guided Process Control Troubleshooting** 117

Appendix A **I.S.A. Standard and Instrument Symbols** 119

Index 131
Bibliography 133

Unit 1

Process Characteristics

MANUAL OBJECTIVE

Introduce you to the dynamics of process control systems, open-loop and closed-loop processes, block diagrams, and various types of processes.

DISCUSSION OUTLINE

The Discussion of Fundamentals covers the following points:

- Process control system
- The study of dynamical systems
- The controller point of view
- Dynamics
- Types of processes
- Process characteristics

DISCUSSION OF FUNDAMENTALS

Process control system

A process is a **dynamical system**, that is, a system that evolves with time. The goal of a process control system is to maintain the process variables as close to their ideal operating values as possible. Figure 1 shows an example of a simple level process. In this level process, water comes in an open vessel at a given flow rate and exits the vessel at another flow rate. A sight glass allows an operator to read the level of water in the vessel. The operator can use a valve to adjust the input flow rate and control the level of water in the tank.

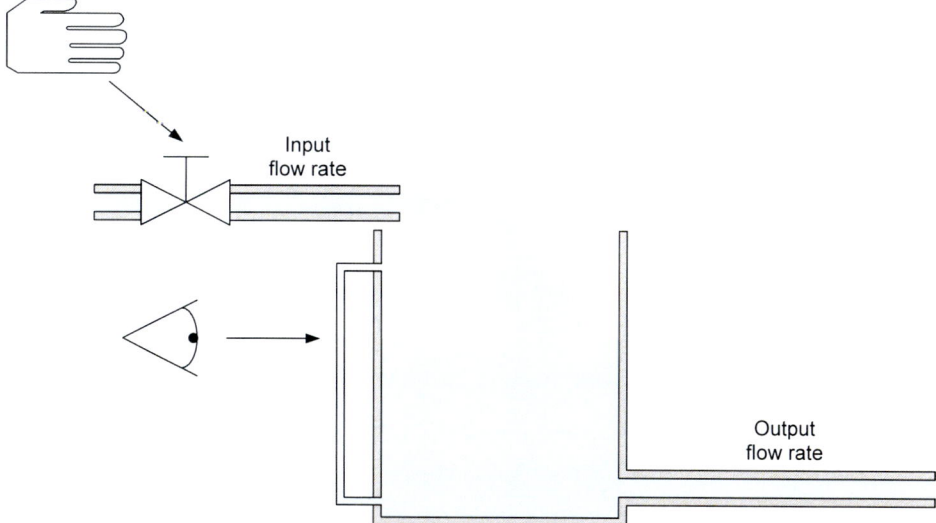

Figure 1. Level process.

Open loop and closed loop

Before getting to the heart of the matter, we must make an important distinction between an open loop and a closed loop. Let us use the level process shown in Figure 1 as reference to illustrate the difference between the two types of loops.

Figure 2 a) illustrates the case where the operator looks elsewhere. In this situation, he can change the opening of the valve but the flow adjustment is not related to the measured variable. A loop where the measured variable has no influence on the manipulated variable is an **open loop**.

Figure 2 b) shows the case where the operator looks at the sight glass and changes the input flow rate as a function of the measured level, the loop is a **closed loop**. In a closed loop, the measured variable has an influence on the manipulated variable. Here, the level (measured variable) helps the operator to adjust the opening of the input valve (manipulated variable) to keep the water in the tank at the desired level (set point). Figure 2 illustrates the two types of loops.

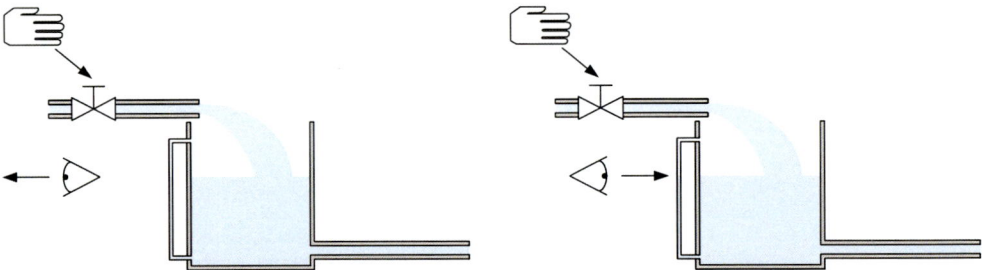

Figure 2. a) Open loop b) Closed loop.

Variables in a process control system

At a given time, the state of a process is a function of the value of different variables. The variables which undergo evolution over time are the ones which are of primary concern to process control. They are: the **controlled variable**, the **manipulated variable**, and the **disturbance**(s). The efficiency of a process depends almost exclusively on the controlled variable. Because the performance of a process relies on the controlled variable, it is crucial to keep it at a desired value; this ideal value is the **set point**. In the level process example of Figure 1, the controlled variable is the level. The operator wants to keep the level below a given value to avoid overflow. The operator can vary the manipulated variable dynamically to keep the controlled variable as close as possible to the set point. He can also compensate for the disturbances that may cause a deviation of the controlled variable from the set point. A variation in the output flow is an example of disturbance.

Operations in a process control system

To counterbalance the disturbances and keep the controlled variable at a given set point, the operator or the equipment must perform various operations. We can group these operations into three main categories: measurement, decision, and correction. In the level process of Figure 1, the operator accomplishes the three operations. In a plant, however, it is more likely that electronic devices take care of these operations, leaving the operator available for the more important tasks of supervision, maintenance, and troubleshooting of the system. Table 1 lists and details the operations in a process control system.

Table 1. Operations in a process control system.

Operation	Devices carrying out the task	Description of the task
Measurement	Sensor, measuring instrument, transmitter	Measure the process variable that needs to be controlled.
Decision	Controller, programmable logic controller (PLC), distributed control system (DCS)	Compare the measured variable to the set point and, based on this comparison, compute the correction to apply to the controlled variable.
Correction	Control valve, actuator, drive	Apply the correction to the controlled variable.

The devices measuring the process variable are usually a sensor/transmitter pair. The sensor is called the **primary element** and the transmitter the **secondary element**. The device that makes a correction as a result of the controller decision is called the **final control element**.

The study of dynamical systems

The study of dynamical systems can be approached from two very different directions. This first is called the internal view. This type of approach attempts to describe how the system works internally. Although it seems a good idea at first to try to understand how a system works and, from that understanding to try to determine how it will react to a disturbance, this can prove to be a laborious process, involving complex mathematics, and it is sometimes an impossible task due to the sensitivity of some systems to initial conditions. An example of a system described internally is the trajectory of an object. The equations describing the trajectory of an object are well known from classical mechanics and, given some initial conditions, one can predict relatively well where the object will land or its position along the trajectory at any given time.

The other way to describe dynamical systems is to view them from an external point. This type of approach comes from electrical engineering. Instead of trying to determine how a system works internally, we only look at the inputs and outputs. Everything in between is like a black box and not useful for an external description. From now on, we will use the external view to describe process control systems. The influence of electrical engineering on process control has been so strong that is has colored most of this field of study. Frequently, we will refer to examples from electricity to give a wider dimension to your understanding of process control.

Block diagrams

Viewing dynamical systems from an external point has several advantages. One of them is the possibility of illustrating their behavior using block diagrams. This type of diagram helps to understand the relationship between the different elements of the system and is frequently used to describe process control systems. Block diagrams usually consist of circles and rectangular boxes. A box in a **block diagram** has one input and one output. The box usually represents an operation such as a multiplication or a division, but it may also represent an important element in the system. A circle represents an addition or a subtraction, thus it has two inputs and one output. Figure 3 shows the block diagram of the level process of Figure 1.

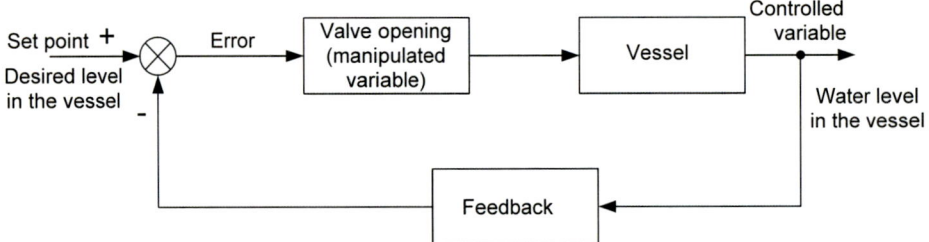

Figure 3. Block diagram of a level process control system.

The controller point of view

Most industrial processes are controlled electronically. The "brain" of such processes is the controller. It is the controller that determines the action to take in order to stabilize the process. Thus, to understand the dynamics of processes, we must look at the information received by the controller.

Figure 4 shows the piping and instrumentation diagram (P&ID) of a level process. The controller in this diagram receives a signal from the level transmitter. For the controller, the transmitter output *is* the controlled variable. From this diagram it is also clear that the controller does not receive information from the control valve. Thus, <u>for the controller</u>, the manipulated variable is not the opening of the control valve, but its own output. Although these distinctions may seem unimportant, keep in mind that they may prove useful for troubleshooting process control systems.

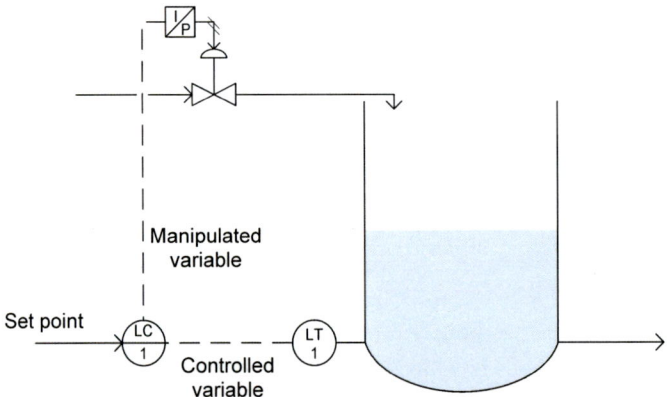

Figure 4. P&ID of a level process.

If we look at the P&ID of the level process again and we circle everything that is before the transmitter output and everything that is after the controller output, what is left? Only the controller! Indeed, from the controller's point of view, the process to control is everything else (Figure 5).

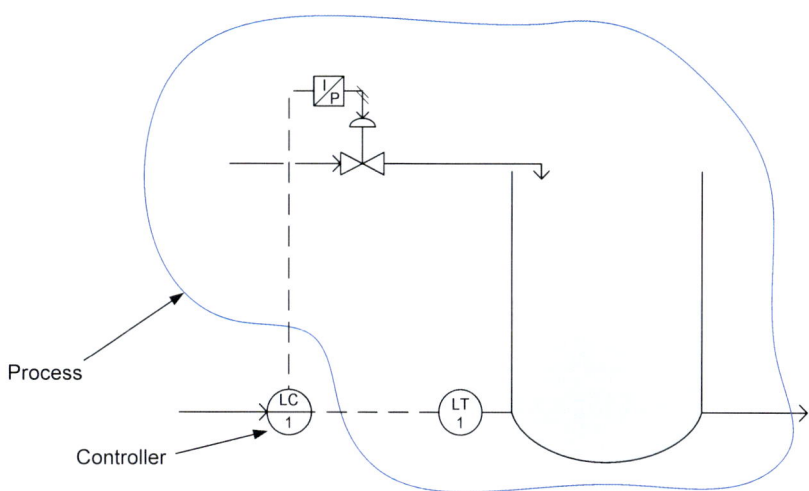

Figure 5. The process as viewed by the controller.

Dynamics

In a dynamical system, actions that the various devices (or the operator) perform do not have an instantaneous effect on the system. In dynamical systems, there is a delay between the action and the effect of this action on the system. Illustrations of this principle can be observed in many situations of everyday life. In your house, turning on the hot water faucet does not provide hot water immediately. Sometimes the water in the pipe has cooled down and you have to wait for hot water to come all the way from the water heater to the faucet. Similarly, when a car driver applies the brakes, the car does not come to a stop immediately.

How dynamical systems respond to an action is determined by many factors. However, as far as process control is concerned, we can divide these factors in three categories: resistance, capacitance, and inertia.

Resistance

Most processes have some **resistive part**. The resistive parts of a system oppose the transfer of energy or mass. An example of **resistance** encountered in a process control system is the pressure loss that pipes and instruments cause. A half-opened valve along a pipe line is a resistance. Changing the load of this resistance (i.e., the valve opening) produces an immediate and proportional change in the flow rate. Reducing the valve opening slows the flow rate of the liquid in the pipe. Processes that consist only of resistive elements are called resistance-type processes or proportional only processes. Figure 6 shows a purely resistive process.

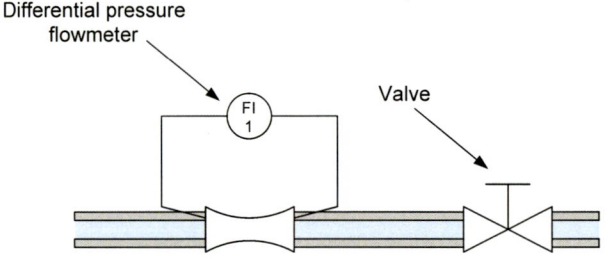

Figure 6. Purely resistive process.

Capacitance

In an electrical circuit, a **capacitance** is an element that can accumulate electrical energy. The capacitor takes a certain amount of time to charge, it stores a given amount of energy as an electric field, and this energy is given back to the system later. Capacitance in a process control system is similar to capacitance in an electrical circuit. It has the ability to store either energy or material. The simplest type of capacitive element in a process control system is a liquid storage tank with an inflow. Figure 7 shows such a tank installation. In this example, the storage tank accumulates water and the level rises at a rate inversly proportional to the tank capacitance. If the resistance in the inflow is neglected, this is an example of a purely capacitive process in terms of level (not flow). Similarly to an electric capacitor, the tank stores water and the rate at which the level rises is inversely proportional to the capacitance of the tank.

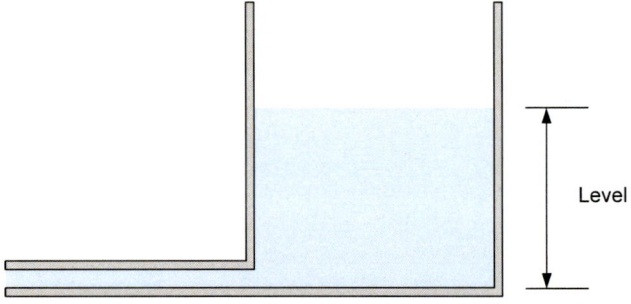

Figure 7. Purely capacitive process.

Inertia

Inertia has a minimal effect on the response of most process control systems with few mechanical parts. However, the effect of inertia is sometimes non-negligible in flow systems that accelerate or decelerate large quantities of fluid.

Types of processes

Process control is concerned with the evolution of systems with time. Therefore, the best way to classify a process is to observe the evolution of the output variable when there is a **step change** in the input variable. By observing the shape of the curve of the output variable as a function of time you can usually determine which type of process it is. Figure 8 shows how you can use a step change to obtain the **response curve** of a process. The shape of the system response to the step change determines the type of process.

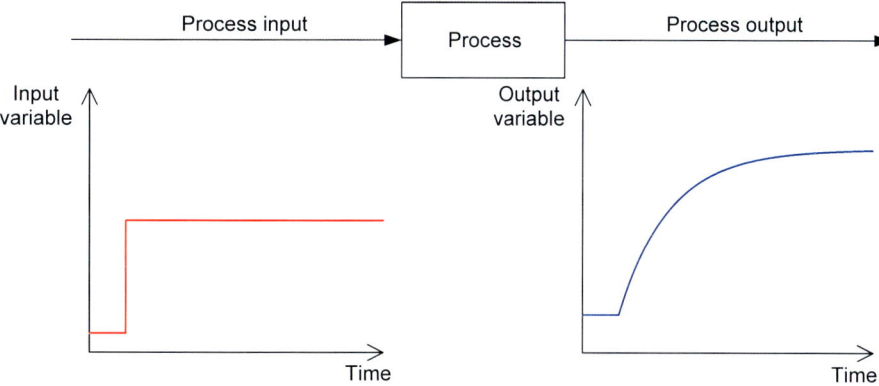

Figure 8. Classifying a process.

There are two main classes of processes: **self-regulating** processes and **non-self-regulating** processes. In a self-regulating process, if the input variable changes, the output variable will stabilize to a new value after a certain period of time. Contrary to self-regulating processes, non-self-regulating processes do not stabilize, if the input variable changes, the output variable will increase (or decrease) linearly or even exponentially without stabilizing. For the moment we will concentrate our efforts on self-regulating processes since they are more common than non-self-regulating processes. Figure 9 compares the response curve of a self-regulating process to the response curve of a non-self-regulating process.

> The pressure control in a boiler is an example of non-self-regulating process. If too much heat is applied, the vapor may not exit fast enough and the pressure inside the boiler will rise instead of reaching a steady state.

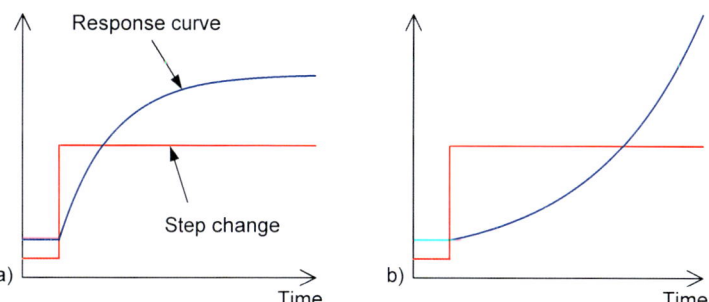

Figure 9. a) Self-regulating process b) Non self-regulating process.

Single-capacitance processes

We have already discussed resistance and capacitance elements in a process. Most of the time, a process is not purely resistive or capacitive. It is common to see a resistance element (e.g., a valve) and a capacitance element (e.g., a vessel) combined in a system. When a resistance element and a capacitive element are combined, the system is an **RC circuit**. In many aspects, the behavior of such an RC circuit is analogue to the behavior of an electric RC circuit made of a resistor and a capacitor.

An RC circuit that has one resistance element and one capacitance element is called a **single-capacitance process**. Single-capacitance processes are self-regulating processes because they tend to stabilize after a step change in the input variable. Single-capacitance processes are also called processes with **first-**

order response because their response curve is the solution to the first-order differential equation that represents the process. Figure 10 shows a single-capacitance process that consists of a vessel with a constant input flow and an output flow that is a function of the opening of a valve. The vessel is the capacitance element since it can accumulate water and the valve is the resistive element. Thus, the input variable of the system is the input flow and the output variable is the level of water in the vessel. The input variable (i.e., the input flow rate) is identified as *m* and the output variable (i.e., the level) is identified as c because in a process control system the input variable is called the manipulated and the output variable is called the controlled variable.

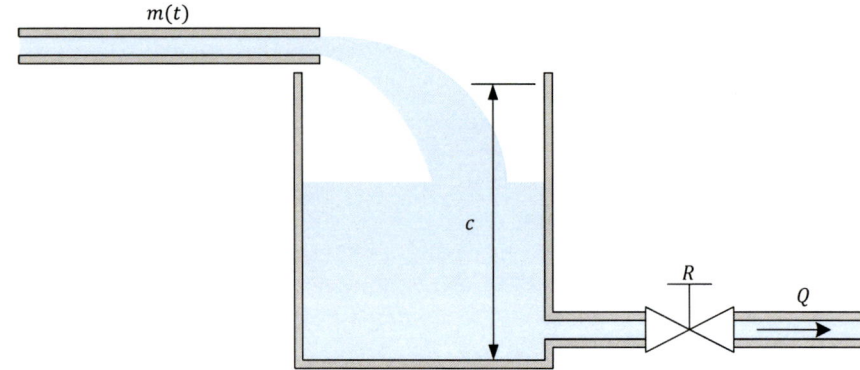

Figure 10. Single-capacitance level process.

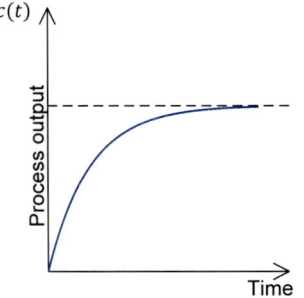

Figure 11. Response curve of a single-capacitance process.

Figure 11 shows the typical shape of the response curve of a single-capacitance process such as the level process presented above. The easiest way to recognize a single-capacitance process is to look at the graph of the output variable as a function of time and determine if the curve has shape similar to the curve at the left.

The mathematics behind single-capacitance processes

This section mathematically describes the single-capacitance process shown in Figure 10. Although this section is not necessary to understand single-capacitance processes, it may help those of you who are familiar with differential equations. Moreover, the information in this section will prove useful to those who wish to study advanced process control.

The **differential equation** modeling single-capacitance processes is:

$$\tau \frac{d}{dt}c(t) + c(t) = Km(t)$$

where τ is the time constant of the process
$c(t)$ is the output of the process
K is the gain (a constant)
$m(t)$ is the input of the process

This equation is a linear first-order differential equation. The differential equation is of the first order because the highest derivative in the equation is of the first order.

For a vessel that is filling, where the input flow is $m(t) = M_0$ for $t > 0$ and $m(t) = 0$ for $t < 0$, the solution to the differential equation modeling single-capacitance processes is:

$$c(t) = KM_0\left(1 - e^{-t/\tau}\right)$$

From this solution we can see why the curve of Figure 11 has such a shape. Indeed, the solution to the differential equation modeling a single-capacitance process is an exponential function.

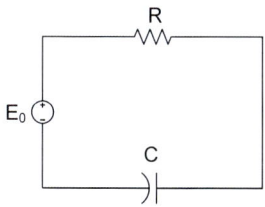

Figure 12. Electrical RC circuit.

The electrical equivalent of the level process presented above is the RC circuit shown in Figure 12. Like its level process equivalent, this RC circuit has one capacitance element and one resistive element, a capacitor and a resistor respectively. Figure 13 a) shows the curve of the voltage across the capacitor as a function of time. This curve is similar to the response curve of the level process discussed above. Figure 13 b) shows the curve of the current in the RC circuit as a function of time. This type of curve is also typical to single-capacitance process. In fact, both curves are exponential curves; precisely they both exhibit an exponential decay behavior because their rate of change is decreasing with time. The curve of Figure 13 a) differs from the second on the position of the asymptote, which is its horizontal limit, and by a constant.

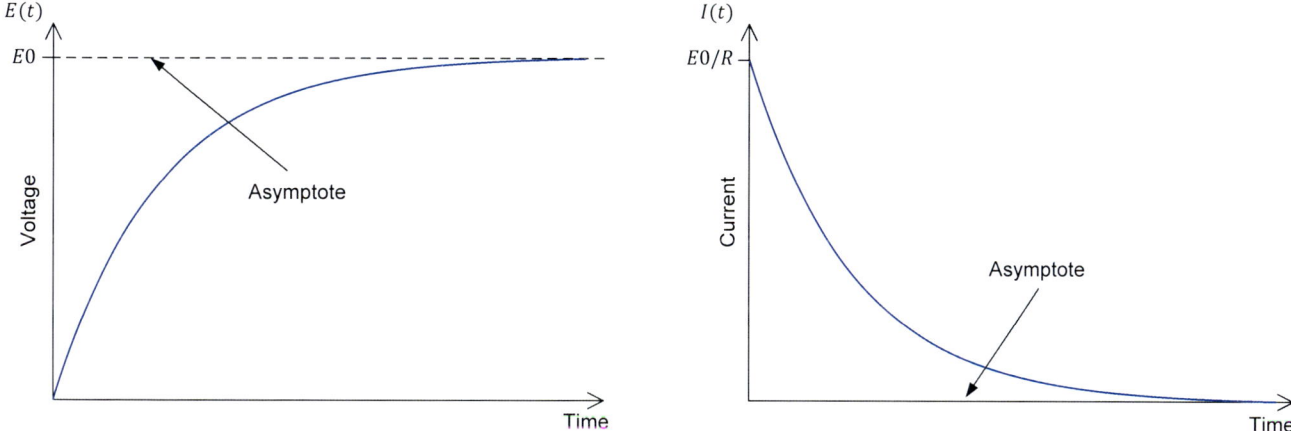

Figure 13. a) Voltage across the capacitor vs. time b) Current in the RC circuit vs. time.

The mathematics behind electrical RC circuits

The current and voltage in the RC circuit of Figure 12 can also be modeled using first-order differential equations. The table below gives the differential equation for the voltage across the capacitor and for the current in the circuit. It also gives the solution to these differential equations.

	Voltage across the capacitor	Current in the circuit
Differential equation	$RC\dfrac{dE}{dt} + E = E_0$	$R\dfrac{dI}{dt} + \dfrac{1}{C}I = 0$
Solution	$E(t) = E_0(1 - e^{-t/(RC)})$	$I(t) = \dfrac{E_0}{R} e^{-t/(RC)}$

Looking at both solutions, it is clear why the curves shown in Figure 13 differ in shape. The solution for the voltage across the capacitor as a function of time has an extra constant E_0 and the exponential function is subtracted from this constant.

If you compare the solution in the table above to the solution to the differential equation modeling single-capacitance processes, you will note that the exponents of these equations look similar. For the level process equation, the time constant is equal to the resistance of the process multiplied by the capacitance (i.e., $\tau = RC$). Similarly, the RC term in the solution to the differential equations of the RC circuit is the capacitive time constant of the circuit. Thus, for both systems, $\tau = RC$.

The time constant is an important characteristic of dynamical processes; it is discussed in detail in the section below.

Process characteristics

The *Measurement* manual for pressure, flow, and level describes the dynamic characteristics of measuring instruments. What applies for measuring instruments, also applies to process control systems. This section describes the dynamical characteristics of a single-capacitance process. Figure 14 shows a first-order response curve for a single-capacitance process. This figure also shows, above the response curve, the step change in the manipulated variable.

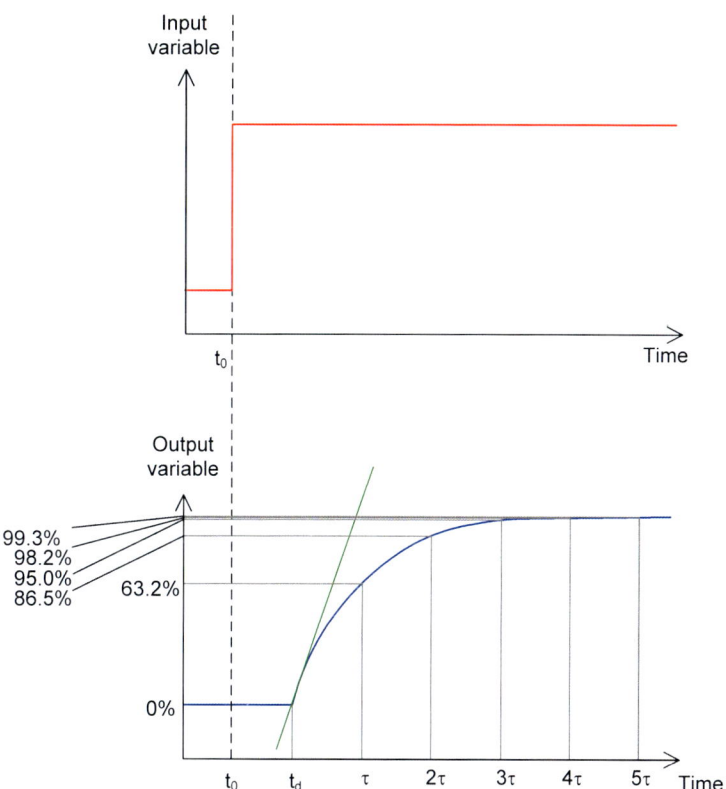

Figure 14. Response curve of a single-capacitance process.

Dead time

The **dead time**, t_d, of a process is the time period elapsing between the step change of the input variable (at t_0) and the first variation of the output variable. During the dead time, nothing changes in the output signal. The time it takes to transport a solid, a liquid, or a gas to the process causes a dead time. Some mechanical or electronic parts also take time to react; this delay is also part of the dead time of the process. The dead time is also influenced by the time the process takes to react (**process lag**), the time the sensor(s) take(s) to react, and the time the controller takes to react (**control lag**). Figure 15 shows an example of process lag. In this example, a candle is heating a metal bar and the heat takes time to reach the hand holding the bar. The time it takes to the heat to reach the hand is the process lag.

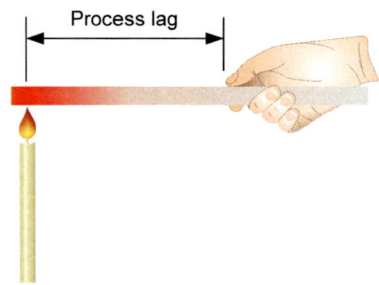

Figure 15. Process lag.

Time constant

The response curve of a single-capacitance process, shown in Figure 14, has a slope that changes with time. The slope of the curve is the rate at which the output variable changes with time. The slope of the response curve is at its maximum when the curve starts to rise. The **time constant**, τ, is the time it would have taken for the output variable to reach its final value if it had kept its initial rate of change. For a process with a first-order response, the time constant corresponds to the time it takes for the output variable to reach 63.2% of the total increase or decrease that follows the step change in the input variable minus the dead time, if any. The time constant depends on the time delay that resistance(s) and capacitance(s) cause. Figure 16 shows the relation between the time constant and the maximum slope of a first-order response curve.

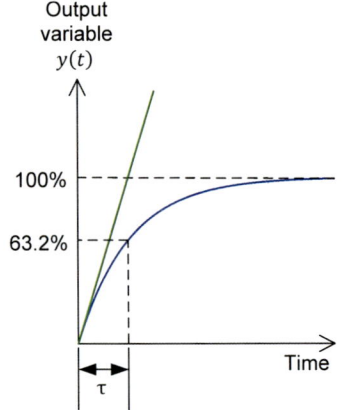

Figure 16. Time constant.

The mathematics behind the time constant

Single-capacitance processes have a response curve that follows the relation below:

$$y(t) = y_0 \left(1 - e^{-t/\tau}\right)$$

where τ is the time constant of the process
       $y(t)$ is the output of the process
       y_0 is maximum value of the output variable
       t is the time

Since this equation is an exponential function, it will reach its maximum at the initial time $t = 0$. The slope of this curve is the first derivative of the function:

$$\frac{dy}{dt} = \frac{y_0}{\tau} e^{-t/\tau}$$

Thus, the slope at $t = 0$ is $\frac{y_0}{\tau}$. The equation of a line with this slope is:

$$z = \frac{y_0}{\tau} t$$

This line reaches the maximum value of the exponential curve y_0 when the time is equal to the time constant τ (i.e., $z = y_0$ at $t = \tau$).

At $t = \tau$ the equation for a single-capacitance process becomes:

$$y(\tau) = y_0(1 - e^{-\tau/\tau})$$

$$y(\tau) = y_0(1 - e^{-1})$$

$$y(\tau) = y_0(1 - 0.368)$$

$$y(\tau) = 0.632 y_0$$

Thus, the time it takes for the process to reach 63.2% of its maximum value is the time constant of the process. Using this method, it is easy to deduce that the process reaches 86.5% of its maximum value at $t = 2\tau$, 95.0% of its maximum value at $t = 3\tau$, 98.2% of its maximum value at $t = 4\tau$, and 99.3% of its maximum value at $t = 5\tau$.

Process gain

The **process gain**, K_p, is the ratio of the output variable change to the change in the input variable. The output variable and input variable are expressed as a percentage of span, thus the process gain is dimensionless.

$$K_p = \frac{\Delta output}{\Delta input} \tag{1-1}$$

Figure 17 illustrates this definition using the graph of the step change in the input variable and the response curve of the process to this step change.

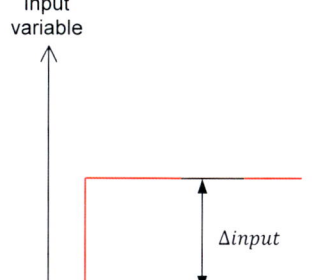
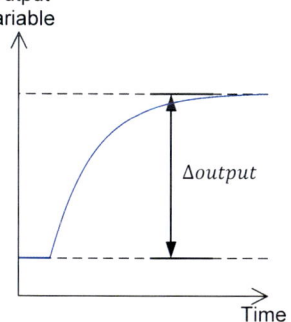

Figure 17. Process gain.

Other characteristics

Steady state	The output reaches 99.3% of its maximum value after five time constants. At this point, you can consider that the process is at **steady state**.
Response time	The **response time** is the time it takes for the process output to reach a predetermined percentage of the steady-state value. In Figure 14 the response time to reach 95% is about $t_d + 3\tau$.
Rise time	The **rise time** is the time it takes for the process output to change from a small to a large percentage of the steady-state value. For example, the rise time of a process can be the time it takes for the process output to go from 5% to 95% of the steady-state value.
Settling time	The **settling time** is the time it takes for the process output to stabilize within a predetermined percentage of the steady-state value. This percentage can be for example, 2%. In the example shown in Figure 14, the process output enters and remains within 2% of the steady-state value after a settling time of about $t_d + 4\tau$.

Exercise 1-1

Determining the Dynamic Characteristics of a Process

EXERCISE OBJECTIVE Familiarize yourself with three methods that allow determining the dynamic characteristics of a process.

DISCUSSION OUTLINE The Discussion of this exercise covers the following points:

- Open-loop method
- How to obtain an open-loop response curve
- Preliminary analysis of the open-loop response curve
- Analyzing the response curve

DISCUSSION

Open-loop method

Ultimately, the purpose of determining the dynamic characteristics of a process is to obtain enough information on the process to be able to tune the controller for efficient process control. There are two different approaches for tuning a controller. The closed-loop approach uses the automatic mode of the controller, while the open-loop approach uses the manual mode of the controller. Throughout this manual you will use open-loop approaches to tune your controller. This type of approach gives a quick estimate of the controller tuning settings.

The method that this section presents will result in an open-loop response curve. In this method, the controller is set to manual mode and is only used to create the step change in the input variable that will trigger a process response. This method requires a self-regulating process. For non-self regulating processes, a different method must be used to tune the controller.

An analysis of the open-loop response curve enables the determination of the following process characteristics:

- The dead time, t_d

- The time constant, τ

- The process gain, K_p

- The order of the response (first order or n^{th} order)

How to obtain an open-loop response curve

To obtain the open-loop response curve of a process, you must have a system with:

- A primary/secondary element that is properly installed and configured
- A recorder with at least two channels
- A controller
- A calibrator
- A final control element

Setting the recorder

To record both the response curve and the step change with the recorder, both the calibrator and the output of the secondary element must be connected to a channel of the recorder. Figure 18 shows a typical setup that allows for the recording of both the response curve and the step change. In this setup, there are two current loops. In the first loop, the calibrator that will produce the step change is the input for channel 1 of the recorder and channel 1 is the input for the final control element. The current loop is closed by connecting the negative connector of the final control element to the negative connector of the calibrator. In the second loop, the output of the secondary element is connected to channel 2 of the recorder, which is then connected to the controller.

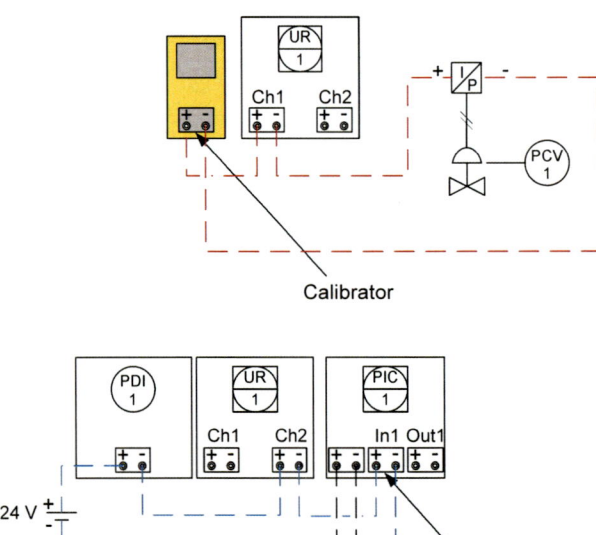

Figure 18. Typical response curve recording setup.

Ex. 1-1 – Determining the Dynamic Characteristics of a Process ♦ *Discussion*

If available, a digital recorder allowing export of the recorded data to a spreadsheet software should be used. Using a spreadsheet software to analyze the response curve gives more precise results than a graphical analysis alone.

Both channels of the recorder should be plotted in units of 0% to 100% of the measured variable range. The calculation to determine the tuning parameters of the controller will be easier if you set the units for the horizontal axis to minutes or fractions of minutes. This can be done directly on the recorder or afterward in the spreadsheet software.

Steps to obtain the response curve

Below are the general steps to obtain the response curve:

1. Make sure your controller is in manual mode.

2. Start your system and set the calibrator output to a given value (e.g. 60%).

3. Wait for the system to stabilize and start recording the calibrator output and the measured variable on the recorder.

4. Create a step change in the manipulated variable by suddenly changing the calibrator output.

5. Wait for the system to be at steady state.

6. Stop your system and prepare your data for analysis.

Preliminary analysis of the open-loop response curve

Determine the process order

Remember that the analysis of the response curve should provide four essential characteristics of the process. One of these characteristics is the order of the process. Before selecting the method of analyzing the response curve, you can determine if your process is a single-capacitance process (first-order) or a multiple capacitance process (n^{th} order) just by looking at the shape of the response curve. Figure 19 shows the difference between the response curve of a single-capacitance process and a multiple-capacitance process. This figure also shows the tangent to the curve at the point where the slope is maximum. The latter has a response curve with an exaggerated "S" shape. On such a curve, the point at which the slope is maximum is in the "S" instead of at the beginning of the curve. This point is the inflection point of the curve, which is the point where the curvature changes sign.

Ex. 1-1 – Determining the Dynamic Characteristics of a Process ♦ *Discussion*

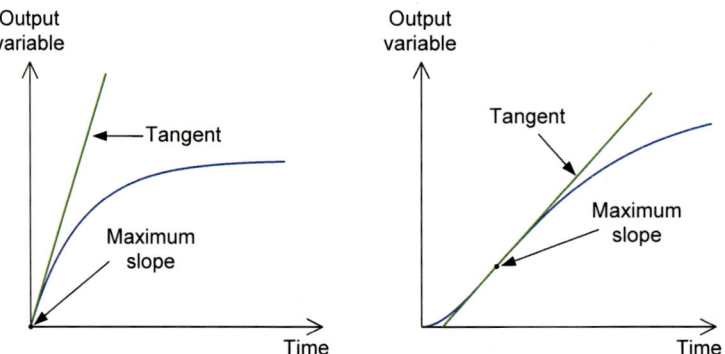

Figure 19. a) Single-capacitance process. b) Multiple-capacitance process.

Once you have determined if your process is a single-capacitance process or a multiple capacitance process and you have calculated the process gain, you must prepare the response curve for further analysis with one of the three suggested methods. This will allow you to determine the dead time and the time constant of the process.

Determine the process gain

You can easily determine the process gain by dividing the percentage of change in the process variable after the step change ($\Delta output$) by the height of the step change in percent ($\Delta input$). Figure 20 shows how you can determine the process gain using the response curve. The gain of the process with the response curve that Figure 20 shows is:

$$K_p = \frac{\Delta output}{\Delta input} = \frac{80\%}{30\%} = 2.67$$

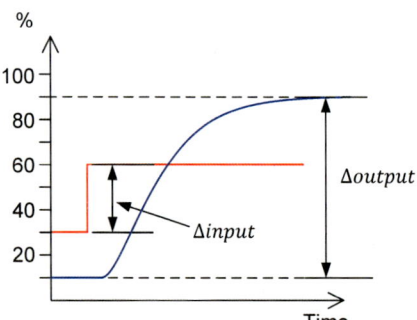

Figure 20. Calculating the process gain.

Prepare the response curve for analysis

A little bit of preparation is required before you can analyze the response curve with one of the methods below. Figure 21 shows a typical response curve before preparation for analysis. On this graph, the response curve starts before the step change and it does not occupy the vertical scale from 0% to 100%.

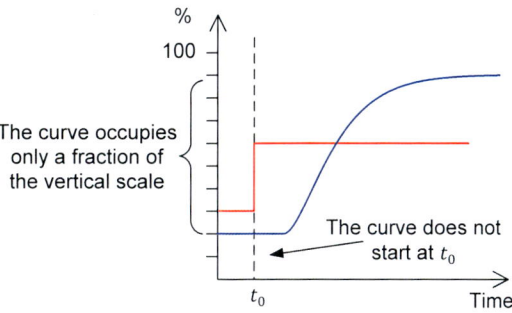

Figure 21. Response curve before preparation for analysis.

To allow an easier analysis it is convenient to plot the data on a new graph, with the horizontal time scale starting at the moment the step change was created. You must also set the vertical scale so that the curve starts at 0% and reaches 100% when it is at steady state. This way, the curve occupies 100% of the vertical scale of the graph.

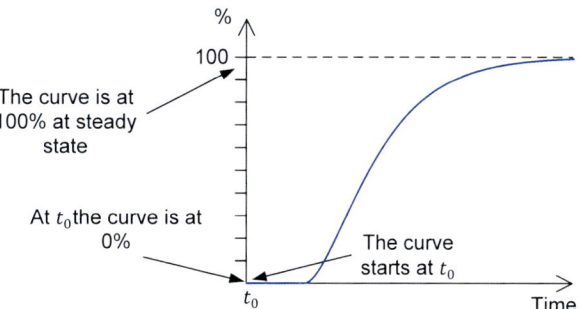

Figure 22. Response curve ready for analysis.

Analyzing the response curve

The approach to determine the gain and the process order from the open loop response curve is straightforward and does not vary from one method to another. However, there are different methods for determining the time constant and the dead time of a process from the open-loop response curve.

This section provides three methods for analyzing the response curve. Although these methods give slightly different results, they are all acceptable and suitable for most processes. The first one is a **graphical method** suggested by Ziegler and Nichols as part of their well known method for tuning PID controllers. This graphical method requires a fine and careful analysis of the graph and may give only middling results. The two other methods give more consistent results since they rely on the analysis of the data rather than the graph.

Graphical method

This method of analysis requires a paper copy of the response curve ready for analysis. On the response curve you must determine the point where the curve is the steepest. For a first-order response curve, this point is right where the curve starts to rise as Figure 23 a) shows. For an nth order response curve the maximum slope is at the inflection point, where the curvature of the response curve changes from concave to convex as Figure 23 b) shows. Once you have determined the point where the slope is at its maximum, draw a tangent line passing through this point.

On the graph, the point where the line intercepts the abscissa is the dead time. For a first-order curve, the dead time is the time elapsed before the process variable starts to rise. For an nth order curve, the process variable begins to change before the dead time ends. The time constant of the process is the time it takes for the process variable to reach 63.2% of its maximum value. For a first-order process, the time constant also corresponds to the point where the line you have drawn intercepts the 100% asymptote.

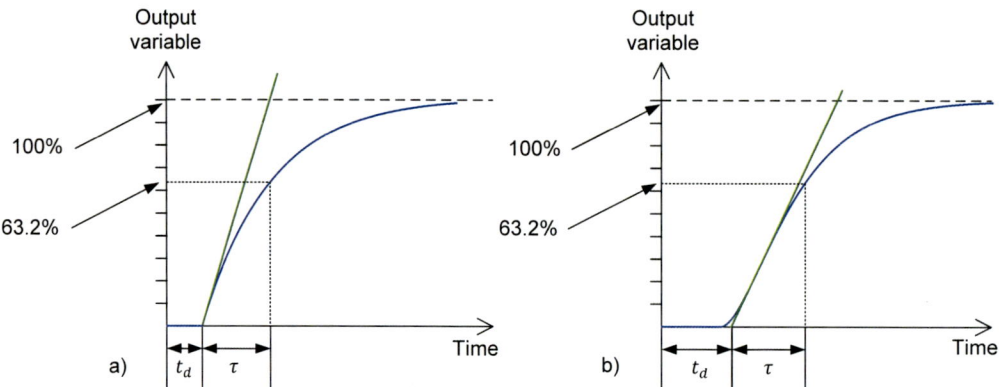

Figure 23. Graphical method.

2%–63.2% method

For n^{th} order response curves, it is sometimes difficult to determine the position of the inflection point. To eliminate error due to the interpretation of the curve, you can use this second method. With this method, the dead time corresponds to the time it takes for the process variable to reach 2% of the total change. The time constant is the time it takes for the process variable to increase from 2% to 63.2%. Figure 24 illustrates this method.

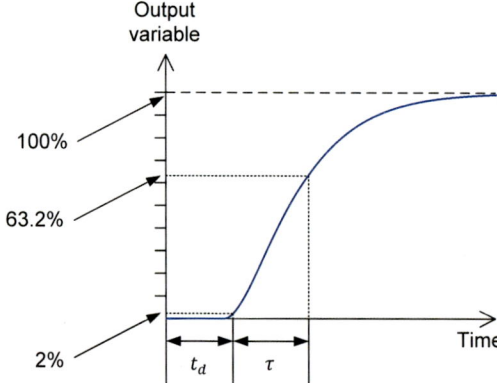

Figure 24. 2%–63.2% method.

28.3%–63.2% method

The third method consists of evaluating the time it takes for the process variable to reach 28.3% and 63.2% of the 100% span. Once you have these two values, use Equation (1-2) to calculate the time constant and Equation (1-3) to calculate the dead time. Figure 25 illustrates this method.

$$\tau = 1.5(t_{63.2\%} - t_{28.3\%}) \tag{1-2}$$

$$t_{d} = t_{63.2\%} - \tau \tag{1-3}$$

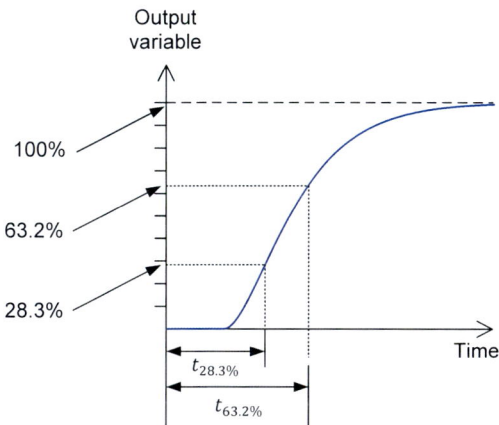

Figure 25. 28.3%–63.2% method.

PROCEDURE OUTLINE

The Procedure is divided into the following sections:

- Set up and connections
- Obtaining the characteristics of a pressure process

PROCEDURE

Set up and connections

> Before trying this exercise you should have successfully completed all the exercises in the *Measurement* manual.

1. Connect the equipment accoring to the piping and instrumentation diagram (P&ID) shown in Figure 26 and use Figure 27 to position the equipment correctly on the frame of the training system. To set up your system for this exercise, start with the basic setup presented in the *Familiarization with the Instrumentation and Process Control Training System* manual and add the equipment listed in Table 2.

Table 2. Material to add to the basic setup for this exercise.

Name	Model	Identification
Differential pressure transmitter (high-pressure range)	46920	PDIT 1
Solenoid valve	46951	S
Electrical unit	46970	
Pneumatic unit	46971	
Accessories	46993	
Calibrator	----	

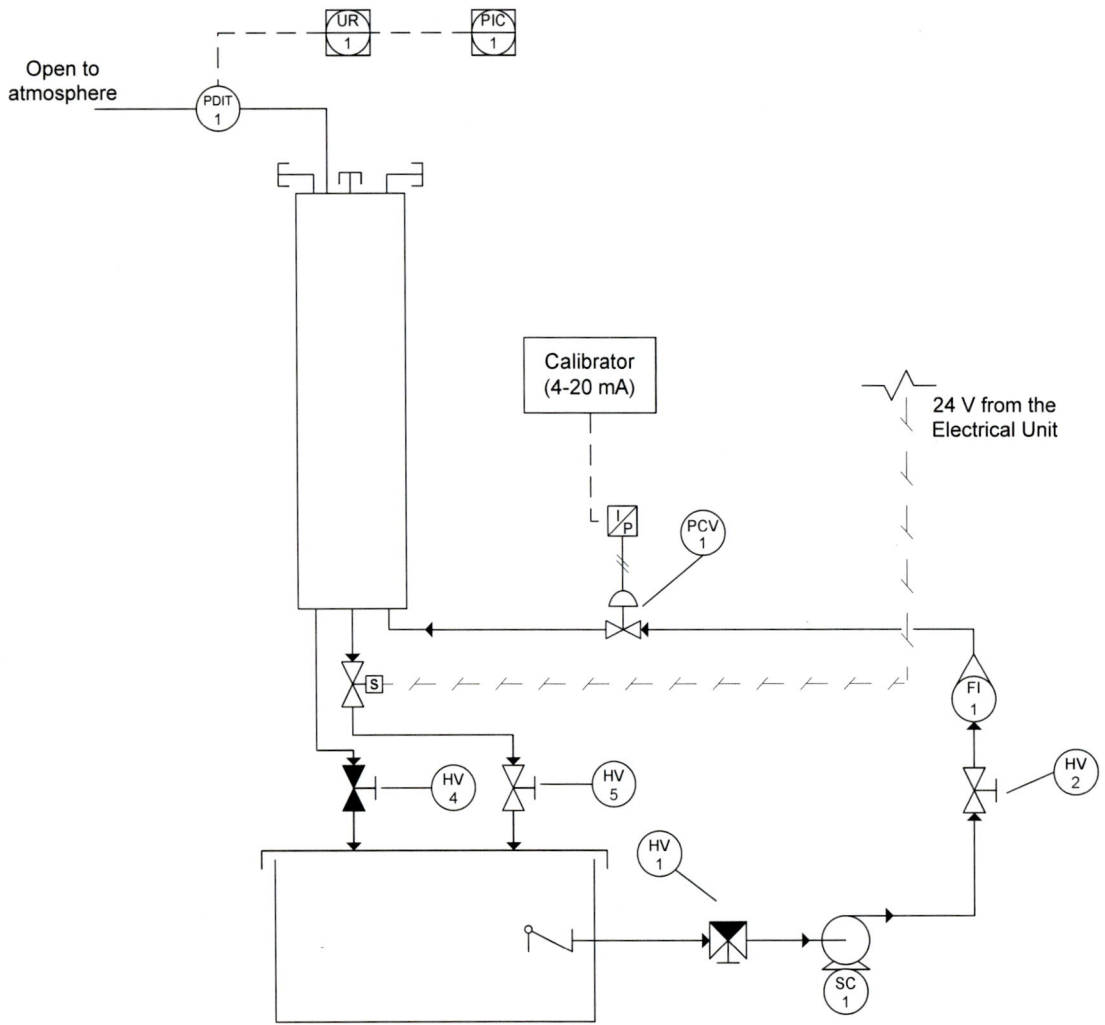

Figure 26. P&ID.

Ex. 1-1 – Determining the Dynamic Characteristics of a Process ♦ *Procedure*

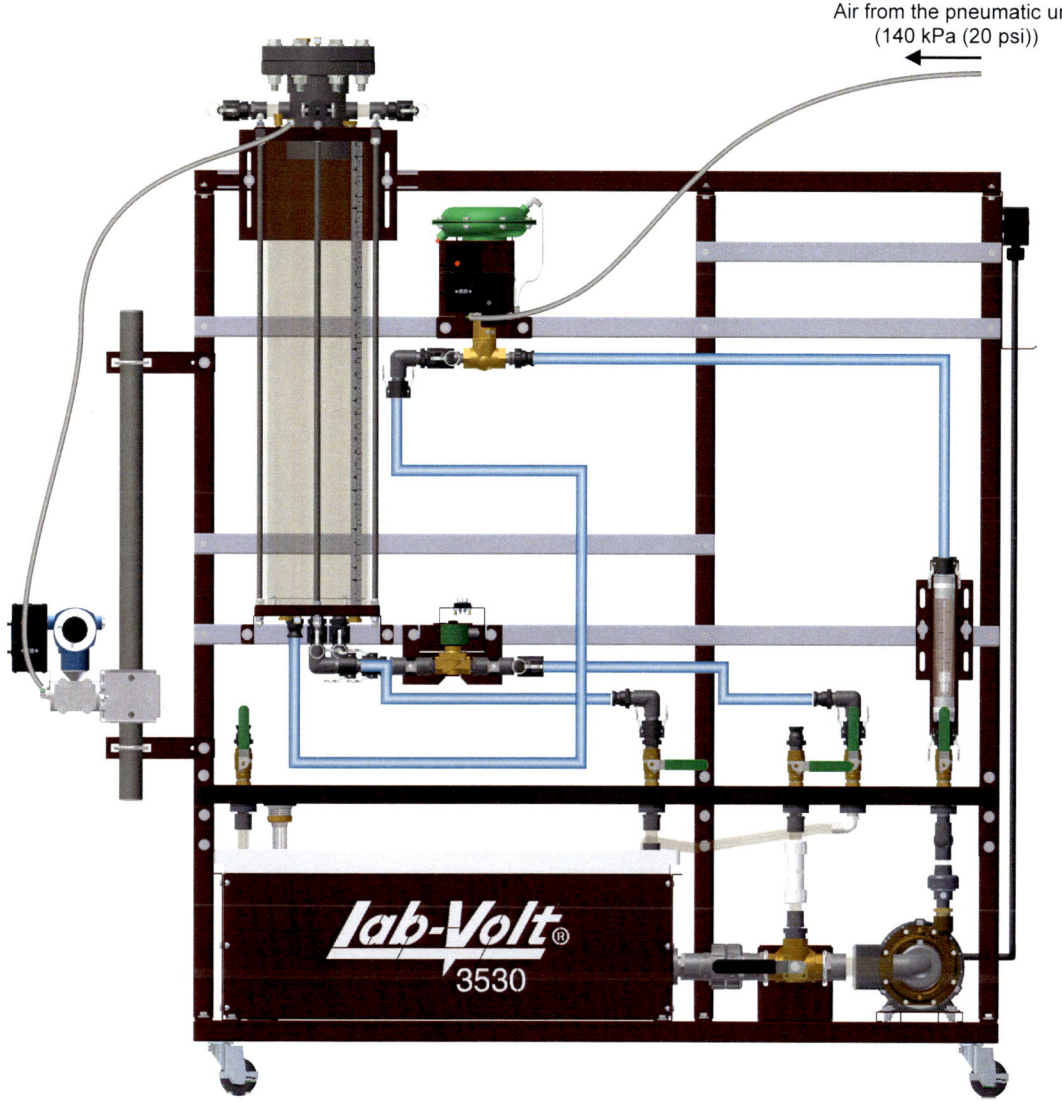

Figure 27. Setup.

2. Connect the control valve to the pneumatic unit. Details about the installation and operation of the control valve are available in the *Familiarization with the Instrumentation and Process Control Training System* manual.

3. Connect the pneumatic unit to a dry-air source with an output pressure of at least 700 kPa (100 psi).

4. Wire the emergency push-button so that you can cut power in case of emergency. The *Familiarization with the Instrumentation and Process Control Training System* manual covers the security issues related to the use of electricity with the system as well as the wiring of the emergency push-button.

5. Do not power up the instrumentation workstation yet. You should not turn the electrical panel on before your instructor has validated your setup—that is not before step 12.

6. Connect the solenoid valve so that a voltage of 24 V dc actuates the solenoid when you turn the power on at step 12.

7. To determine the dynamic characteristics of your process, you must connect your calibrator to the control valve and the differential pressure transmitter to the input of the controller. You must include the recorder in your connection. On channel 1 of the recorder, plot the signal from the calibrator and on channel 2, plot the signal from the transmitter. Be sure to use the analog input of your controller to connect the differential pressure transmitter. Refer to the manual of your controller for details on how to connect it to other devices.

8. Figure 28 shows how to connect the paperless recorder to your system to plot the calibrator signal on channel 1 and the controller input on channel 2.

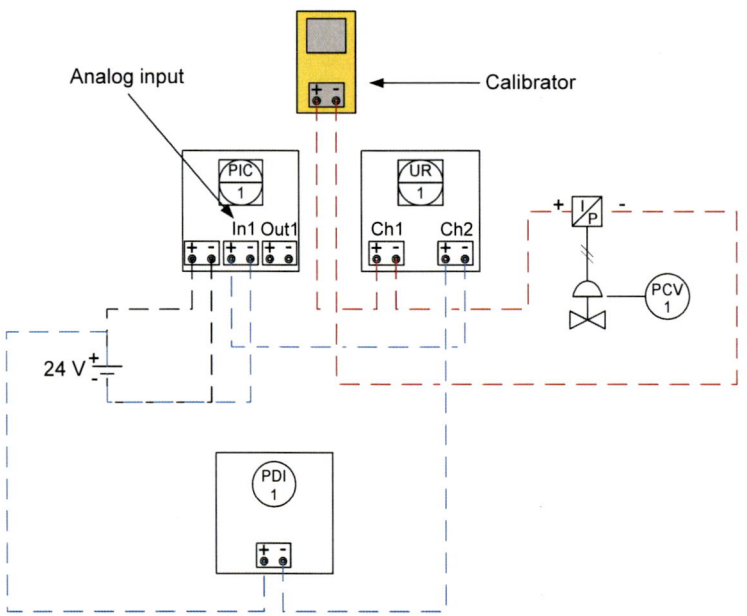

Figure 28. **Connecting the equipment to the recorder.**

9. Before proceeding further, complete the following checklist to make sure you have set up the system properly. The points on this checklist are crucial elements to the proper completion of this exercise. This checklist is not exhaustive, so be sure to follow the instructions in the *Familiarization with the Instrumentation and Process Control Training System* manual as well.

- ☑ All unused male adapters on the column are capped and the flange is properly tightened.
- ☑ The solenoid valve under the column is wired so that the valve opens when the system is turned on.
- ☑ The ball valves are in the positions shown in the P&ID.
- ☑ The three-way valve at the suction of the pump (HV1) is set so that the flow is directed toward the pump inlet.
- ☑ The control valve is fully open.
- ☑ The pneumatic connections are correct.
- ☑ The controller is properly connected to the differential pressure transmitter.
- ☑ The calibrator is properly connected to the control valve.
- ☑ The paperless recorder is connected correctly to plot the calibrator signal on channel 1 and the controller input on channel 2.

10. Ask your instructor to check and approve your setup.

11. Remove one of the caps from the top of the column. This maintains the pressure in the column at atmospheric pressure.

12. Power up the electrical unit, this starts all electrical devices as well as the pneumatic devices.

13. Use the calibrator to send a 4 mA signal to the current to pressure converter of the control valve. When the converter receives a 4 mA signal, a 20 kPa (3 psig) signal is sent to the control valve. At this pressure signal, the valve is fully open.

14. Test your system for leaks. Use the drive to make the pump run at low speed to produce a small flow rate. Gradually increase the flow rate, up to 50% of the maximum flow rate that the pumping unit can deliver (i.e., set the drive speed to 30 Hz). Repair any leaks.

15. Fill the pipes completely with water.

Obtaining the characteristics of a pressure process

16. Make sure the impulse line of the differential pressure transmitter is free of water and that it is connected to the pressure port at the top of the column.

17. Configure the differential pressure transmitter so that it gives pressure readings in the desired units. Set transmitter parameters so that it sends

a 4 mA signal if the pressure is 0 kPa (0 psi) and a 20 mA signal if the pressure is 80 kPa (11.6 psi). Refer to the *Familiarization with the Instrumentation and Process Control Training System* manual for details on the configuration of the differential pressure transmitter.

18. Adjust the zero of the differential pressure transmitter. The column is at atmospheric pressure because of the removed cap; therefore the transmitter will read 0 kPa (0 psi) when the pressure inside the column is equal to the atmospheric pressure.

19. Replace the column connector cap removed at step 11. This will allow pressure to build in the column when you turn the pump on.

20. Send a 12 mA signal to the control valve using the calibrator. This half opens the control valve. If the paperless recorder is correctly configured, channel 1 of the recorder should indicate that the calibrator output is 50%.

21. Set the pump to its maximum speed and wait for the pressure reading to stabilize.

Be sure to use the differential pressure transmitter, Model 46920-00. This differential pressure transmitter has a high-pressure range.

22. To create a step change in the process input, set the value of the calibrator signal to 10 mA.

23. On the paperless recorder, watch the change in the value of the process variable when the calibrator output changes from 50% to about 38%.

24. Wait for the value of the process variable to stabilize.

25. Once the system is at steady state, you have all the information to determine the dynamic characteristics of the process.

26. Stop the system.

27. Follow the procedure in the *Familiarization with the Instrumentation and Process Control Training System* manual to transfer the data from the paperless recorder to a computer.

28. Plot the data using a spreadsheet software.

Ex. 1-1 – Determining the Dynamic Characteristics of a Process ◆ *Conclusion*

29. Analyze the data using the three methods presented in this exercise and fill in Table 3 with the results.

Table 3. Characteristics of the process.

	Graphical Method	2%–63.2% method	28.3%–63.2% method
τ (s)			
t_d (s)			
K_p			

CONCLUSION

In this exercise, you have learned three methods of determining the dynamic characteristics of a process. You have used these methods to determine the dynamic characteristics of a pressure process.

REVIEW QUESTIONS

1. What is the order of the pressure process you have analyzed in this exercise? How did you deduce that?

   _____
   _____
   _____

2. How does a process with a large gain react to a step change?

   _____
   _____
   _____

3. Describe what an inflection point for the response curve of an nth order process is.

   _____
   _____
   _____

4. Which of the three methods of analyzing a response curve is most subject to interpretation?

   _____
   _____

5. Which process characteristics does a careful analysis of the open-loop response curve allow you yo determine?

___
___
___

Unit Test

1. A process is

   a. a dynamical system.
   b. a system that evolves with time.
   c. a sequence of actions.
   d. All of the above.

2. Which element can execute the decision task in a process control system?

   a. Primary element
   b. Pump
   c. PLC
   d. None of the above.

3. The pressure loss that pipes cause in a process control system are a type of

   a. resistance.
   b. capacitance.
   c. inertia.
   d. inductance.

4. A single-capacitance process is a system that has

   a. an element of resistance and an element of inertia.
   b. an element of resistance and an element of capacitance.
   c. an element of capacitance and an element of inertia.
   d. None of the above.

5. The response curve of a single-capacitance process has a

   a. parabolic shape.
   b. exponential shape.
   c. hyperbolic shape.
   d. symmetric shape.

6. The response curve of an nth order process has a

   a. C shape.
   b. J shape.
   c. U shape.
   d. S shape.

7. A single-capacitance process has a

   a. first-order response curve.
   b. second-order response curve.
   c. n^{th}-order response curve.
   d. $\hbar$-order response curve.

8. A non-self-regulating process is a

   a. process with a substance that helps regulation.
   b. process that stabilizes slowly.
   c. process that does not stabilize.
   d. None of the above.

9. For a given process, it takes 1 minute 46 seconds for the process variable to go from 28.3% to 63.2% of the span after a step change. What is the time constant for this process?

   a. The time constant for this process cannot be calculated from this information.
   b. 159 seconds
   c. 1 minute 59 seconds
   d. 106 seconds

10. The dead time of a process is 6 minutes and the time constant is 48 minutes. When does the process variable reach 63.2% of the span?

    a. After 48 minutes.
    b. After 42 minutes.
    c. After 54 minutes.
    d. Never.

Unit 2

Feedback Control

MANUAL OBJECTIVE

Learn the basics of feedback control and familiarize yourself with the different types of feedback control, such as: on-off, P, PI, and PID control.

DISCUSSION OUTLINE

The Discussion of Fundamentals covers the following points:

- Feedback control
- On-Off control
- PID control
- Proportional controller
- Proportional and integral controller
- Proportional, integral, and derivative controller
- Proportional and derivative controller
- Comparison between the P, PI, and PID control
- The proportional, integral, and derivative action
- Structure of controllers

DISCUSSION OF FUNDAMENTALS

Feedback control

Feedback control is the type of control that you are most likely to find in industry since it is a very simple control strategy. In feedback control loops, the measurement is made at the output of the process and the process variable has time to change before the transmitter detects anything. If the controlled variable drops below the set point, the controller raises the manipulated variable and, in response to this change, the controlled variable raises above the set point. When the controller receives a signal from the transmitter that indicates that the controlled variable is above the set point, the controller adjusts its output to reduce the error. This causes the controlled variable to drop below the set point, which again triggers the controller to raise its output.

The controller reacts like this each time the controlled variable passes below or above the set point. This produces an oscillatory response that stabilizes after a time. This **oscillatory response** is typical of a feedback control loop.

Simplicity is the main advantage of feedback control. Feedback control also compensates for all disturbances affecting the controlled variable. The main disadvantage of feedback control is that it compensates for disturbances only after the controlled variable has changed and deviated from the set point. The disturbance propagates through the entire process before the system can compensate for the deviation.

In a feedback control loop, the controller tries to maintain the controlled variable as close as possible to the set point. To decode whether it should act on the manipulated variable or not, the controller solves an equation that takes into account the difference between the set point and the controlled variable. This difference is the **error**:

$$e(t) = \mp(r - c(t)) \qquad (2\text{-}1)$$

where $e(t)$ is the error
   r is the set point
   $c(t)$ is the controlled variable

The sign in front of Equation (2-1) depends on the controller action (see section below). If the controller is set to direct action, the sign is negative and if it is set to reverse action the sign is positive.

Once the error is computed, the controller algorithm establishes a mathematical relationship between the error and the controller output. This relationship can be an on-off relationship, a proportional relationship, a derivative relationship, or an integral relationship. It can also be a combination of the last three. This unit details the relationships between the error and the output of the controller that are common in industry. Some controllers have a simple control algorithm while others can be set to different control modes.

Reverse vs. direct action

Before going further, an important feature of controllers must be defined. This is the controller action. A controller can be set either to **direct action** or **reverse action**. If a controller is set to direct action, the controller output signal increases if the controlled variable increases. If the output of the controller is connected to a fail-close control valve (FC), the percentage of opening of the control valve increases as the value of the controlled variable increases. An example of a control loop with a controller set to direct action is the level control loop that Figure 29 shows. With such an installation, if the level in the tank increases, the output of the controller also increases, and the control valve opens to allow liquid to exit from the tank.

> A fail-close valve is a valve that closes on control signal failure (or air failure). Such a valve is fully open if it receives a 100% signal from the controller.

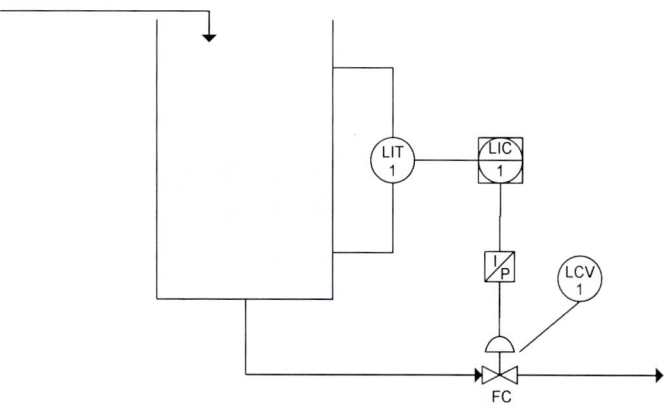

Figure 29. Level control loop with a fail-close control valve.

A controller set to reverse action works the other way around. That is, the output of the controller decreases if the controlled variable increases. Figure 30 shows a level control loop similar to the control loop of Figure 29 with the exception of the control valve which is a fail-open control valve (FO). In this case, if the level in the tank increases, the controller must decrease its output to open the control valve and allow liquid to exit from the tank. Thus, the controller must be set to reverse action.

> A fail-open valve is a valve that opens on control signal failure (or air failure). Such a valve is fully closed if it receives a 100% signal from the controller.

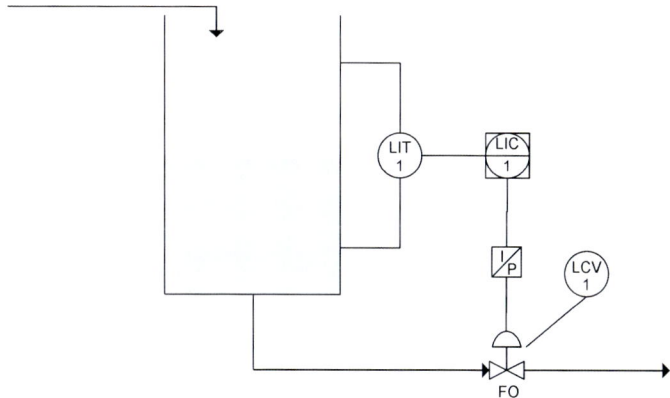

Figure 30. Level control loop with a fail-open valve.

Setting your controller action correctly is essential to controlling a process. The controller action depends on the relationship between the controlled variable and the desired controller output. If the controller action is not set properly, no control can be achieved.

On-Off control

The oldest, least expensive, and simplest type of controller is the on-off or two-position controller. This type of controller is a discontinuous controller because the control signal can take only certain discrete values, usually on and off (0% and 100% of the output signal).

On-off controllers are most effective on large capacitance, slowly changing processes. They are commonly used in home heating and air conditioning systems. They can also be found in appliances such as refrigerators, freezers, and water heaters.

Figure 31 shows the typical input-output relationship of an on-off controller in direct-action mode. When the controlled variable is below the set point, the output signal is at its minimum value (0%), this is the *off* state. When the controlled variable is above the set point, the output signal is at its maximum value (100%), this is the *on* state. The action of this controller is said to be direct because the controller output signal increases (passes from 0% to 100%) as the controlled variable increases.

> Remember that for a controller in a process control loop, the manipulated variable, that is the controller output, is the input variable of the process. Similarly, the controlled variable, that is the transmitter output, is the output variable of the process.

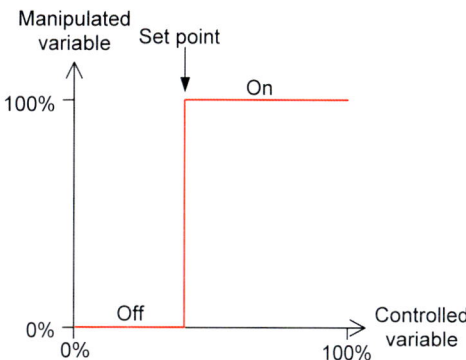

Figure 31. Typical response of an on-off controller in direct-action mode.

Figure 32 shows the typical input-output relationship of an on-off controller in reverse-action mode. The controller output signal is at its maximum value (100%) when the controlled variable is below the set point and passes to its minimum value (0%) when the controlled variable is above the set point. The action of this controller is said to be reverse because the controller output signal decreases (passes from 100% to 0%) as the controlled variable increases. The on-off controller with reverse action is commonly used in home heating systems. When the temperature falls below the set point, the home heating system turns on.

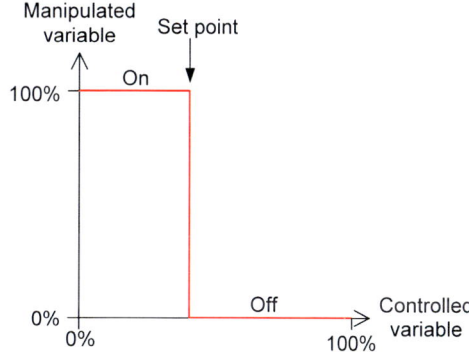

Figure 32. Typical response of an on-off controller in reverse-action mode.

Figure 33 illustrates the response of an on-off controller in reverse-action mode. The controller output signal changes state every time the controlled variable crosses the set point. Thus, the value of the controlled variable oscillates continuously around the set point. The amplitude and frequency of the oscillations is directly related to the process capacitance. The amplitude is high and the frequency is low when the process capacitance is large and vice versa. This explains why on-off controllers are best suited for slow response processes which normally have large capacitances. Moreover, the oscillating nature of on-off control tends to wear on control valves and contactors, which makes this type of control less suitable for industrial continuous processes.

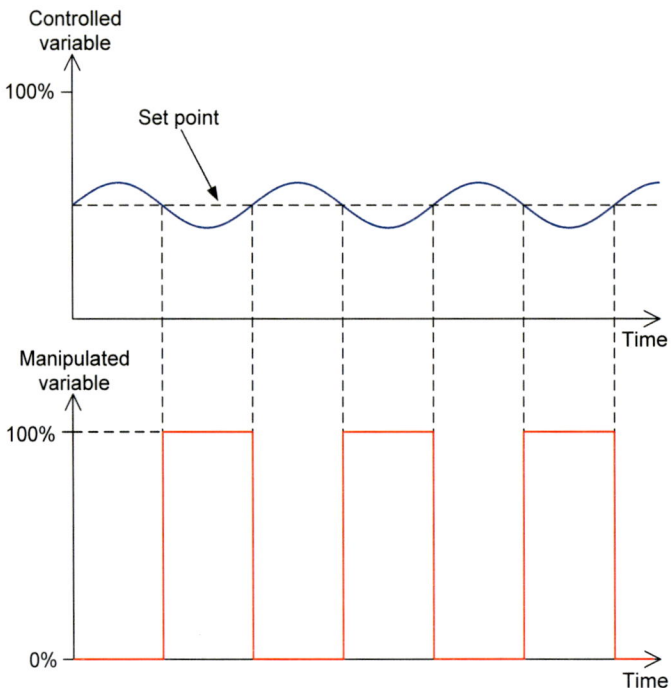

Figure 33. Typical response of an on-off controller in reverse-action mode.

On-off controller with a dead band

The oscillation of the controlled variable around the set point in process control systems using an on-off controller may result in potential problems when the oscillation frequency becomes too high. This may cause the equipment to wear out prematurely, especially the control element which is continuously switched on and off. One way of minimizing this drawback is to reduce the oscillation frequency by adding a dead band (sometimes called differential gap). The **dead band** is a zone around the set point where no control action is taken. The state of the controller changes only if the controlled variable is above (or below) the set point by at least half the value of the dead band.

Figure 34 and Figure 35 show the typical input-output relationships of direct action and reverse action for on-off controllers with a dead band. The controller output signal changes state at two different values located on either side of the set point. The controlled variable must pass through the entire zone covered by the dead band before the controller output signal changes state. As a result, a different path is taken on the input-output relationship depending upon whether the controller output signal passes from the *on* state to the *off* state or from the *off* state to the + state. This phenomenon is referred to as **hysteresis**.

Unit 2 – Feedback Control ♦ *Discussion of Fundamentals*

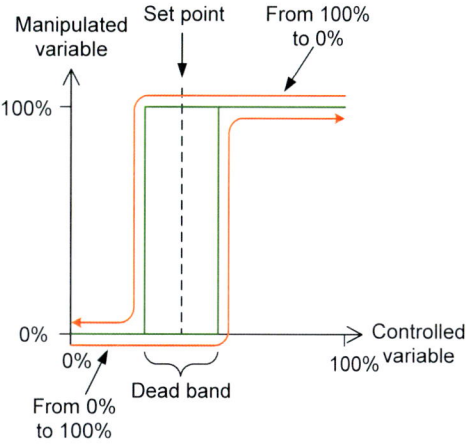

Figure 34. Typical response of an on-off controller in direct-action mode with a dead band.

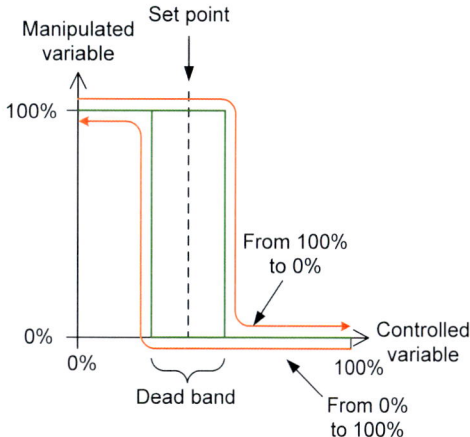

Figure 35. Typical response of an on-off controller in reverse-action mode with a dead band.

Figure 36 shows the response of an on-off controller set to reverse action when a dead band is present. Compared to the response of a controller without dead band, the oscillation frequency is smaller. This effect can be used to prevent cycling at an excessive rate in the system, therefore reducing the wear on control valves. However, the larger the dead time, the higher the amplitude of the oscillation, which means that the controlled variable will go farther from the set point.

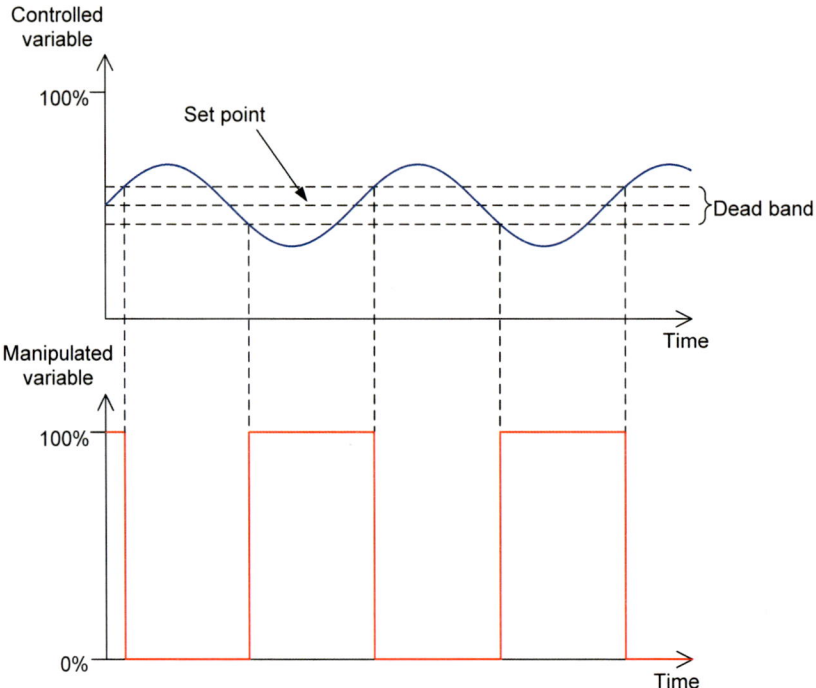

Figure 36. Typical response of an on-off controller in reverse-action mode and with a dead band.

Some devices can only achieve on-off control. However, PID controllers can be set to act like an on-off controller by using a large value for the parameter called the controller gain, K_c.

The **controller gain** determines the magnitude of the proportional action and is the reciprocal of the **proportional band**, PB, as Equation (2-2) shows. The proportional band is a parameter that some controllers use instead of the controller gain. Be careful, the controller gain is <u>not</u> the process gain. However, as we will see later, the process gain allows estimating the optimum controller gain for a given process.

$$K_c = \frac{100\%}{PB\%} \qquad (2\text{-}2)$$

> On-off control is similar to proportional control with a very large controller gain. If the proportional band or a proportional controller is set to zero, the controller gain will be very large (infinite). Therefore, the proportional controller will act as an on-off controller.

PID control

Controllers using a proportional, derivative, or integral relationship between the error and the output are used in more than 95% of the process control applications. A controller that can do proportional (P), integral (I), and derivative (D) control is called a **PID controller**. The most common settings for PID controllers are:

- P (proportional mode only)
- PI (proportional and integral) which is the most used mode
- PD (proportional and derivative) which is seldom used
- PID (proportional, integral, and derivative)

PID control takes its name from the algorithm processing the error to determine the action of the controller. This algorithm has three important terms. The first term is the **proportional** term which is proportional to the error. The second term is the **integral** term which is proportional to the integral of the error and the last term is the **derivative** term which is proportional to the derivative of the error. Table 4 gives the relationship between the manipulated variable, $m(t)$, and the error for each term of the PID algorithm.

Table 4. Proportional, integral, and derivative controller equation.

Relationship with the error	Term
Proportional	$m(t) = K_c e(t) + b$
Integral	$m(t) = \dfrac{K_c}{T_i} \int e(t)\, dt + b$
Derivative	$m(t) = K_c T_d \dfrac{d}{dt} e(t) + b$

The form of the PID control algorithm differs from one controller to another depending on the manufacturer. Equation (2-3) shows the theoretical version of the PID controller algorithm.

$$m(t) = K_c \left(e(t) + \frac{1}{T_i} \int e(t)\, dt + T_d \frac{d}{dt} e(t) \right) + b \qquad (2\text{-}3)$$

where $m(t)$ is the output of the controller (i.e., the manipulated variable)
K_c is the controller gain
$e(t)$ is the error
T_i is the integral time constant
T_d is the derivative time constant
b is the bias

From Equation (2-3) we can see that the output of the controller depends on the three aforementioned terms, in particular the three parameters. These three parameters are the controller gain (K_c), the **integral time** constant (T_i), and the **derivative time** constant (T_d). The influence of these parameters on the way the controller reacts to a deviation from the set point will be thoroughly discussed in the next section. In order to configure a controller to efficiently control a process loop, you must determine the optimal values for K_c, T_i, and T_d for the process. In the following section we will analyze the common controller settings (P, PI, PD, PID).

Proportional controller

A controller running in proportional mode produces an output proportional to the error. Running a controller in proportional mode is the equivalent of running a controller with the full PID algorithm where the value for the integral time is very large and the derivative time is set to zero. You can verify this by setting $T_i = \infty$ and $T_d = 0$ in Equation (2-3):

$$m(t) = K_c \left(e(t) + \frac{1}{\infty} \int e(t)\, dt + 0 \cdot \frac{d}{dt} e(t) \right) + b$$

$$m(t) = K_c e(t) + b \tag{2-4}$$

Equation (2-4) is the general equation for a proportional controller. This is simply the equation of a line with a slope equal to the controller gain. The y-intercept of this line is the bias of the controller. The **bias** is the output of the transmitter when the error is zero. Figure 37 shows the relation between the error and the output for a controller in proportional mode.

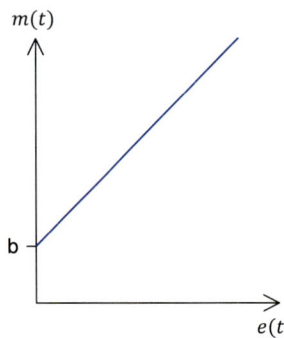

Figure 37. Relation between the error and the output for a proportional controller.

As Figure 37 shows, the greater the error, the stronger the response of the controller. The value of the controller gain determines the sensitivity of the controller to an error. The gain determines the strength of the response of the controller to an error. For a given error, a large gain leads to a stronger controller response than a small gain. If the gain of a proportional controller is very large, the controller acts as an on-off controller.

> The offset left by a proportional controller is also called a steady state error.

The principal disadvantage of a proportional controller is that it returns the controlled variable to steady state in case of perturbation, but always with an offset from the set point. That is, the controlled variable returns to stability, but a little bit above or below the set point. A larger gain helps to reduce this offset; however a large controller gain also produces a more oscillatory response and compromises the stability of the process. Most processes have a gain value above which the process is unstable (i.e. the process oscillates in an increasing-amplitude fashion). This maximum gain value is called the ultimate gain, K_u. Figure 38 shows the controlled variable as a function of time for a process controlled with a proportional controller. In this figure you can observe that for a small gain the offset is large and that for a large gain the offset is smaller but the system is less stable.

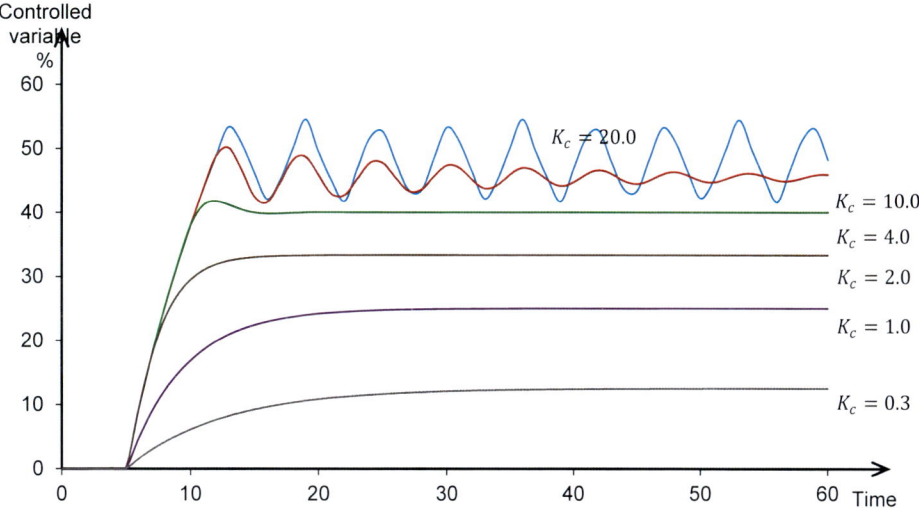

Figure 38. Proportional control with different gain values.

Figure 39 shows the block diagram of the algorithm of a proportional controller.

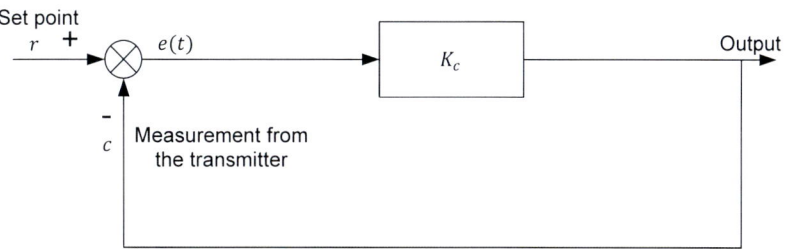

Figure 39. Block diagram of the algorithm of a proportional controller.

Figure 40 shows the typical response of a controller in proportional mode to a set-point change.

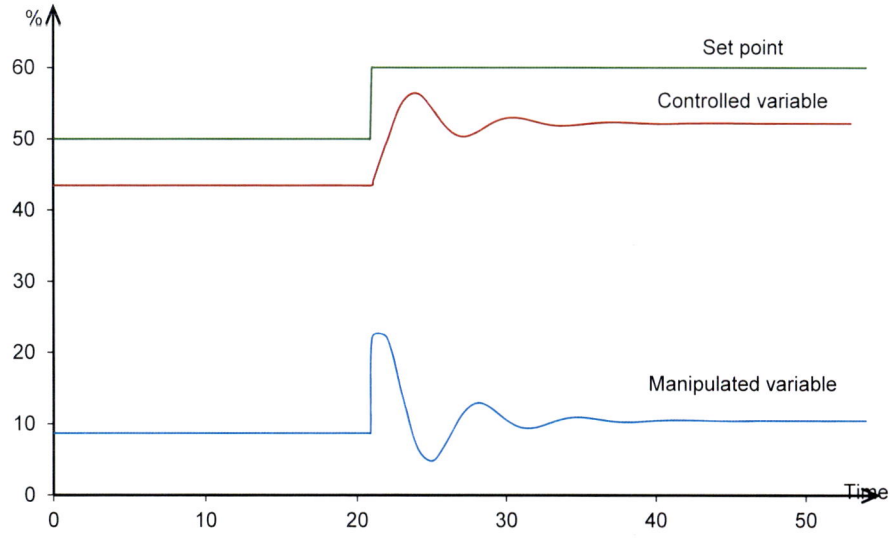

Figure 40. Proportional control.

Tuning a controller for proportional control

As Figure 38 shows, the response of the controller depends on the controller gain. Thus, the only parameter that seems necessary to configure a controller for proportional control is the controller gain. Looking back at Equation (2-4) we can deduce that, with the controller gain and the bias (if required) one can configure a controller for proportional control. However, some brands of controllers do not use the controller gain as a tuning parameter. Instead, they use the proportional band presented in Equation (2-2). Be sure you identify which of these two parameters your controller uses as setting parameters before trying to configure it.

Proportional and integral controller

One disadvantage of proportional control is that the steady state value has an offset from the set point. For many processes this is unfortunate since their ideal operation point is at the set point, not at the set point plus (or minus) an offset. The proportional and integral mode, also known as the PI mode, combines the proportional term and the integral term of the general PID control algorithm. Adding an integral term to the proportional term helps to eliminate the error that remains after the proportional correction.

> The integral mode is also called the reset action.

Running a controller in PI mode reduces the permanent error between the steady state value and the set point. However, the PI mode has a drawback; it leads to a less stable control loop with a longer period of oscillation. The general equation for a PI controller is:

$$m(t) = K_c e(t) + \frac{K_c}{T_i} \int e(t)\, dt + b \qquad (2\text{-}5)$$

The influence of the integral term

The integral mode is always used with the proportional mode (to eliminate the offset left by the proportional term). It is an improvement to the proportional mode; it gives the controller more *intelligence* if it may be called so. The action of the integral term is not instantaneous; it acts over a period of time. The controller starts the integral action when the proportional action acts and the integral term changes continuously unless the error is zero. When the error is zero the integral term remains stable.

As Equation (2-5) shows, the integral term is proportional to the controller gain divided by the integral time constant. The integral time (or reset time) is the time it takes for the integral term to repeat the action of the proportional term. Thus, the integral action increases (or decreases) the output of the controller with time. The velocity at which the controller changes integral action is proportional to the error. The graphs shown in Figure 41 illustrate how the integral term reacts to a positive error, a negative error, and a null error. The two graphs at the left show that if there is a positive error, the controller output increases constantly. The two graphs in the middle show that if there is a negative error, the controller output decreases proportionally to the error. Finally, the two graphs at the right show that if the error falls to zero, the controller output remains constant. Figure 42 shows how the integral bloc of a PI controller responds to a series of errors.

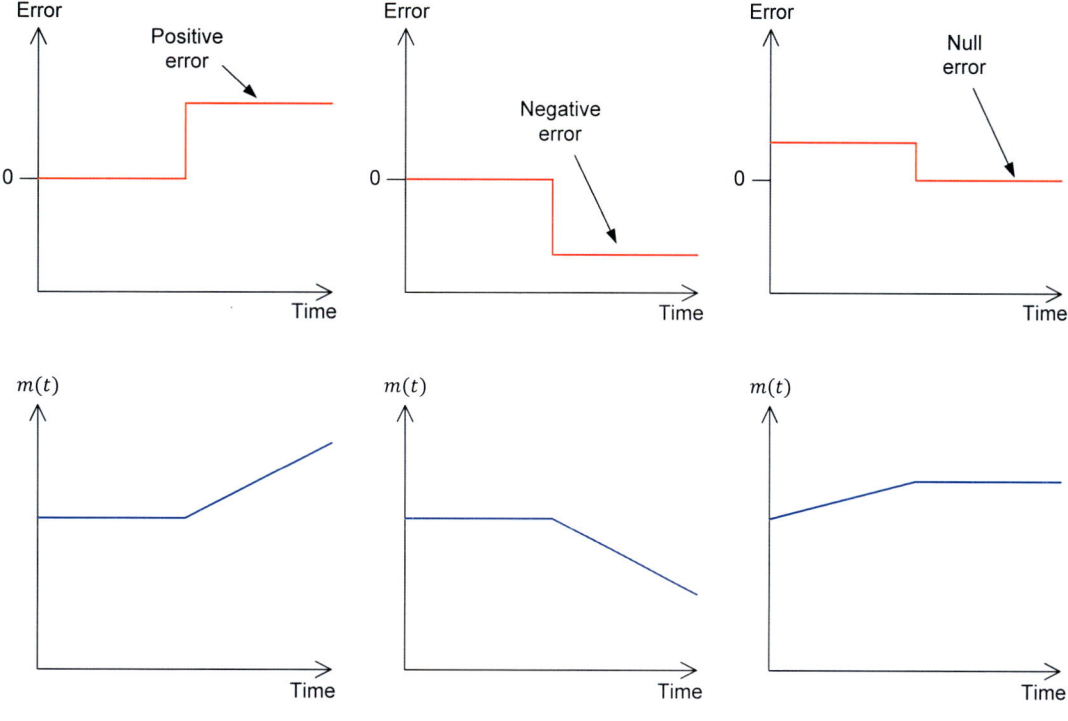

Figure 41. The integral term for different types of errors.

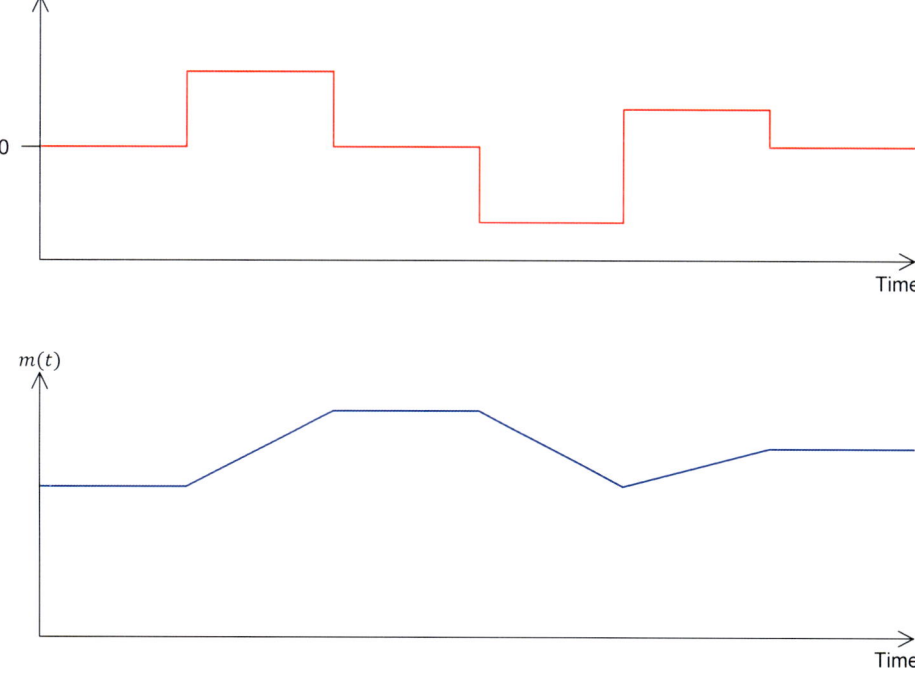

Figure 42. The controller action for different errors.

Figure 43 shows that after a time equal to the integral time constant, the increase in the output of the integral bloc equals the error. If the error remains the same, the output of the integral bloc will be two times the error after a time equal to two times the integral time constant.

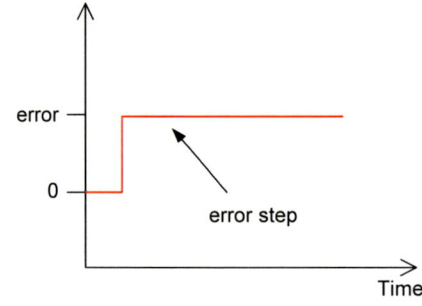

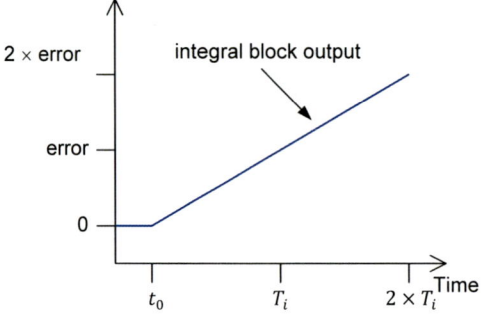

Figure 43. Integral block output as a function of time.

Tuning a controller for PI control

Three parameters are required to tune a controller for PI control, the controller gain (or the proportional band), the bias, and the integral time constant. Like the controller gain, some brands of controllers do not accept the integral time as a parameter. Instead, they use the reciprocal of the integral time constant called the **reset rate**. The reset rate has units of repeats per second (or per minute). Equation (2-6) shows the relationship between the integral time and the reset rate.

$$reset\ rate = \frac{1}{T_i} \tag{2-6}$$

The integral in the integral term

For those of you who are unfamiliar with integration, the following fact about integrals and their use in process control may help you to understand the integral action of a controller.

- The integral of a function $f(x)$ calculated on an interval $[a, b]$ is called a definite integral and is noted $F(x) = \int_a^b f(x)dx$.

- The integral of a function is the area under the curve of this function.

- The area under the curve is negative if it is below the x-axis and positive if it is above the x-axis.

- The integral is the inverse of the derivative. That is $F(x)$ is the function whose derivative is $f(x)$.

The above facts are not strict definitions; they are merely an explanation of integration.

The integral action of a controller is the integral of the error (plus the bias):

$$m(t) = \frac{K_c}{T_i} \int e(t)\,dt + b$$

The action is always calculated over a period of time, thus it is a definite integral. The controller does not read data continuously, it takes readings at a determined frequency called the **scan time**. Therefore, the controller approximates the area under the curve using a small rectangle with a width equal to the scan time. The following figure illustrates this principle using an error curve with a sinusoidal shape.

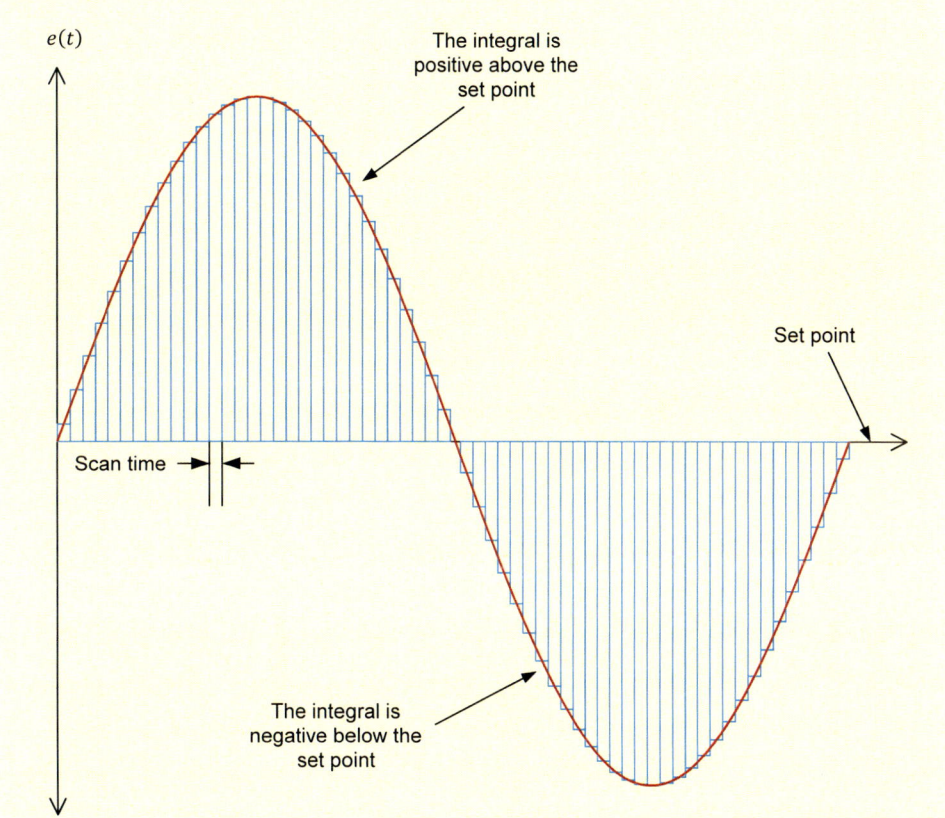

Proportional, integral, and derivative controller

A controller set for PID control uses the full capability of the algorithm. In addition to the proportional term and the integral term presented in the sections above, the PID algorithm includes a derivative term. The derivative mode of a PID controller is sometimes called the **rate action** or the **pre-act**. The general equation for a PID controller is:

$$m(t) = K_c e(t) + \frac{K_c}{T_i} \int e(t)\, dt + K_c T_d \frac{d}{dt} e(t) + b \qquad (2\text{-}7)$$

It is sometimes said that the derivative action anticipates the error. But in fact, the derivative action calculates the rate of change of the error. The rate of change of the error (i.e., its speed) allows the controller to evaluate the trend and act to prevent unwanted changes in the controlled variable. If the derivative of the error is large, the controller knows that the controlled variable is rapidly going away from the set point and it can further modify its output to bring back the controlled variable toward the set point more rapidly. Furthermore, even if the proportional and integral terms are large because the controlled variable is far from the set point, and if the derivative of the error is negative, the controller knows that the error is decreasing and it can reduce its output instead of increasing it further. Overall, the derivative term helps to stabilize the process. It compensates for the instability that the integral term may cause.

On the other hand, the derivative term of the PID algorithm is sensitive to noise. In a process with **noise**, the controlled variable displays rapid variations of small amplitude on the process response curve. These variations occur rapidly and high speed variations produce large derivative values. This causes oscillations in the controller output due to the derivative term. For this reason, PID control is not recommended for fast changing processes such as a pressure process. Any process with a small time constant is susceptible to noise and not suitable for PID control. The control of slow processes however, can be improved using PID control. Most temperature processes are slow processes with a large time constant.

Tuning a controller for PID control

Four parameters are required to tune a controller for PID control: the controller gain (or the proportional band), the integral time constant (or the reset rate), the derivative time, and bias.

Proportional and derivative controller

This control mode is seldom used. It combines the proportional and derivative modes. Because there is no integral term in this mode, the controller will return the controlled variable to stability, but with an offset relative to the set point. Also, because of the derivative term, this control mode is not recommended for processes with noise. A slow temperature process, without noise, which must not be kept exactly at the set point, can be controlled using a PD controller. PD controllers are also used for batch pH control. In this case the derivative term may improve the control compared to proportional control only. Below is the equation for PD control:

$$m(t) = K_c e(t) + K_c T_d \frac{d}{dt} e(t) + b \qquad (2\text{-}8)$$

Comparison between the P, PI, and PID control

Figure 44 compares the response of a first-order system to a step change when there is no action from the controller, a proportional action only, a proportional and integral action, and a proportional, integral, and derivative action. In this figure, you can observe that, when the controller is set in proportional mode only, the controlled variable reaches steady state quickly, but with a permanent offset from the set point. If an integral action is added to the proportional action, the controlled variable reaches the steady state at the set point, but with less stability. Finally, adding a derivative action to the two other reduces the instability introduced by the integral action.

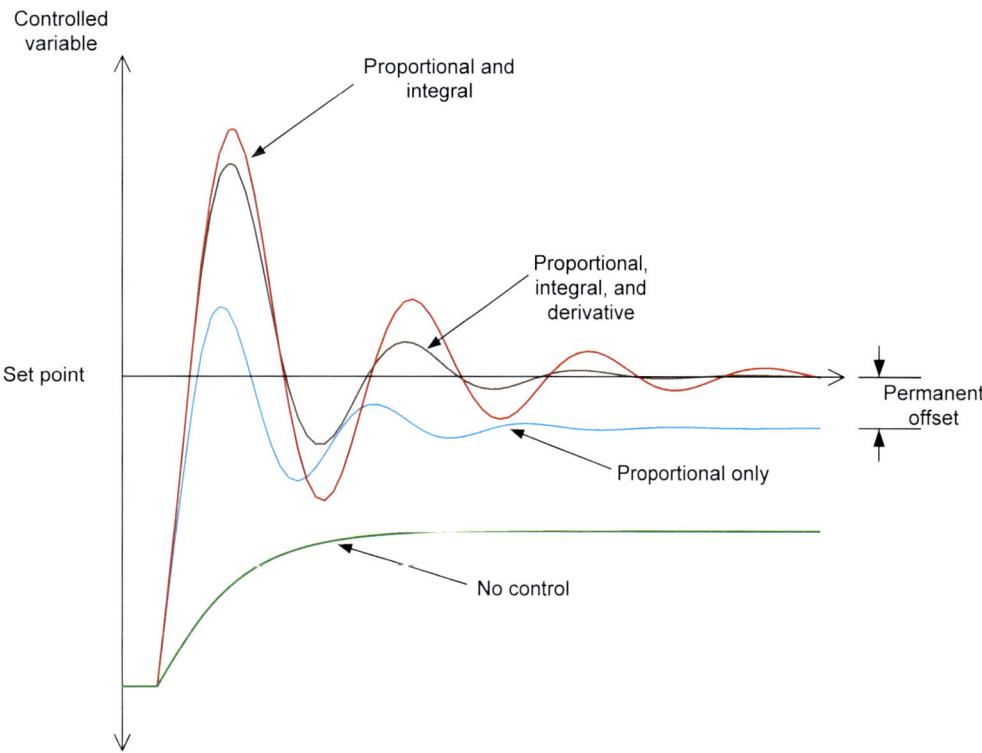

Figure 44. Comparison between the P, PI, and PID control.

The proportional, integral, and derivative action

As seen earlier, the proportional, integral, and derivative actions of a PID controller produce different effects on the controlled variable. The present section shows graphs of the contribution of each type of action to the controller output for a given perturbation.

Figure 45 shows the perturbation measured by the transmitter. This perturbation will be used in this illustration to calculate the error, the proportional action, the integral action, and the derivative action. Carefully observe the different curves and try to understand how each action varies relatively to the error. If you are familiar with the mathematics used in the PID algorithm, referring to Table 4 may help you to see how the shape of each curve is linked to the PID algorithm.

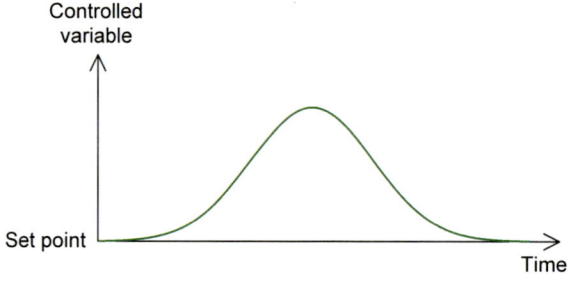

Figure 45. Perturbation.

On the graph above, the set point is on the x-axis and, since the perturbation is just above the set point, the error is simply the negative of the perturbation curve because in this case $e(t) = 0 - c(t) = -c(t)$. Figure 46 shows the error for the perturbation shown above.

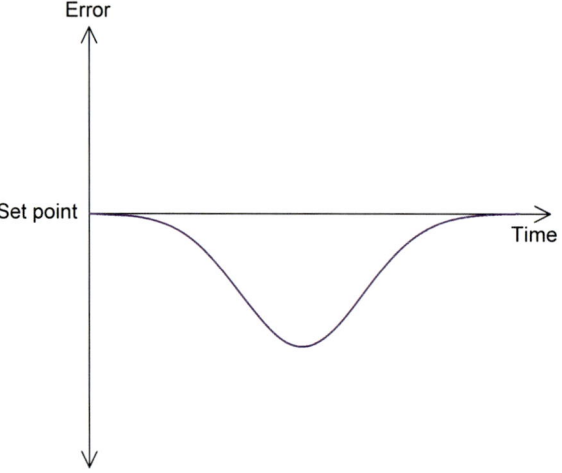

Figure 46. Error.

The proportional action is the error multiplied by the controller gain. Thus, as Figure 47 shows, the proportional action has the same shape as the error curve but a different amplitude.

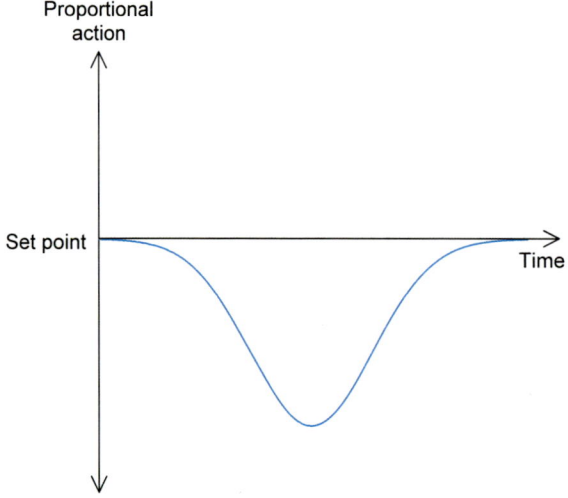

Figure 47. Proportional action.

Figure 48 shows the integral action. The integral action is the area between the error curve and the x-axis as a function of time. At first there is no error, thus the integral action is zero. The integral action increases as the area between the curve and the x-axis increases. When the error is zero again the area stops increasing, therefore the integral action remains constant.

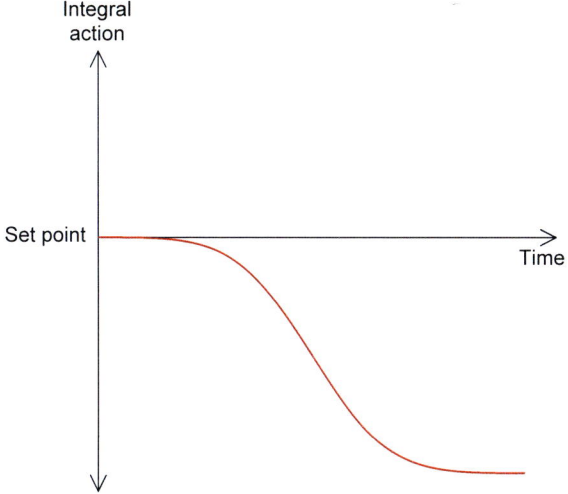

Figure 48. Integral action.

The derivative action is the slope of the error curve as a function of time. You can observe in Figure 49 that there is a peak in the derivative action when the slope of the error curve is maximum. The derivative action is zero when the slope of the error curve is null, which is when the error is at its maximum.

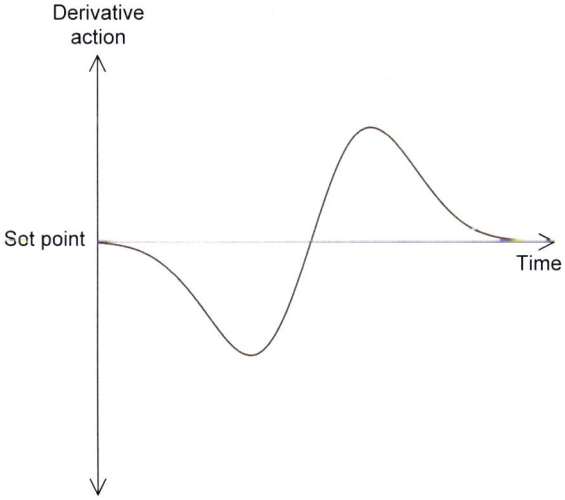

Figure 49. Derivative action.

Structure of controllers

Non-interacting

The algorithms presented in the sections above are from an **ideal controller**, also called **non-interacting controller**. Figure 50 shows the standard structure for a non-interacting controller. The equation for this type of controller is reproduced below. Most controllers do not use this type of configuration for their algorithms because it is too sensitive to noise in PD or PID mode. However it is a convenient model to teach process control since it is simple to understand.

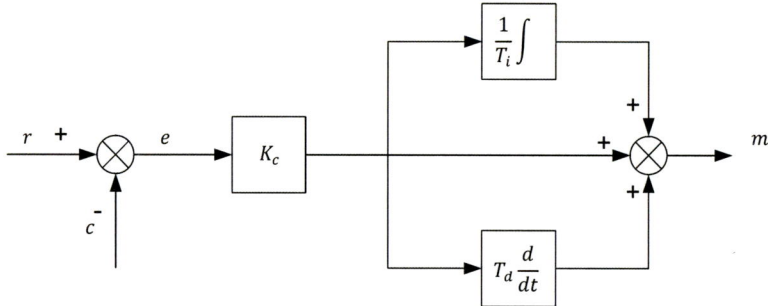

Figure 50. Non-interacting controller.

$$m(t) = K_c \left(e(t) + \frac{1}{T_i} \int e(t)\, dt + T_d \frac{d}{dt} e(t) \right) + b \qquad (2\text{-}9)$$

Interacting

The **interacting controller** (or **serial controller**) is the most common type of industrial controller. As the block diagram of Figure 51 shows, the different terms of the algorithm have an influence on each other. Equation (2-10) shows that the terms do not add to each other. The output is the product of the three blocks ($P \times (I+1) \times (D+1)$). A controller with such an algorithm is much easier to tune and less sensitive to process noise.

The constants K_c', T_i', and T_d' are used to illustrate the difference between the constants of an interacting controller and the constants of a non-interacting controller (K_c, T_i, and T_D).

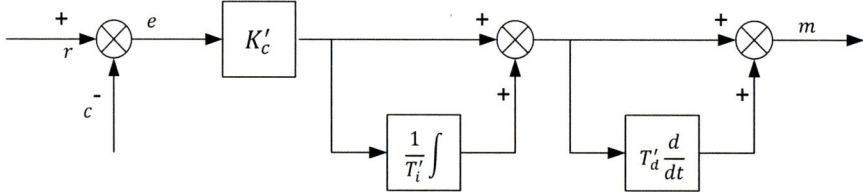

Figure 51. Interacting controller.

Equation (2-10) shows the algorithm for an interacting controller. Multiplying and rearranging the terms of this equation gives Equation (2-11). This equation shows why this algorithm is said to be interacting, because changing T_i', or T_d' affects all three terms (proportional, integral, and derivative).

$$m(t) = K_c' \left(e(t) + \frac{1}{T_i'} \int e(t)\, dt \right) \left(1 + T_d' \frac{de(t)}{dt} \right) + b \qquad (2\text{-}10)$$

$$m(t) = K_c' \frac{T_i' + T_d'}{T_i'} \left(e(t) + \frac{1}{T_i' + T_d'} \int e(t)\, dt + \frac{T_i' T_d'}{T_i' + T_d'} \frac{de(t)}{dt} \right) + b \qquad (2\text{-}11)$$

The mathematical link between the non-interacting and the interacting algorithm

By carefully analyzing Equation (2-11) one may note that, aside from the constants, the equation for an interacting controller is identical to the equation for a non-interacting controller. An interacting controller represented by Equation (2-11) corresponds to a non-interacting controller whose coefficients are:

$$K_c = K'_c \frac{T'_i + T'_d}{T'_i}$$

$$T_i = T'_i + T'_d$$

$$T_D = \frac{T'_i T'_d}{T'_i + T'_d}$$

where K'_c, T'_i, and T'_d are the constants for the interacting controller and K_c, T_i, and T_D are the constants for the non-interacting controller.

Parallel

Another type of controller structure, mainly used for teaching, but also used in PLCs and DCSs, is the parallel structure that Figure 52 shows. The difference between a non-interacting controller and a **parallel controller** is that in a non-interacting controller, the controller gain (K_c) is distributed to all three terms. In a parallel controller, the gain only applies to the proportional term (see Equation (2-12) au-dessous).

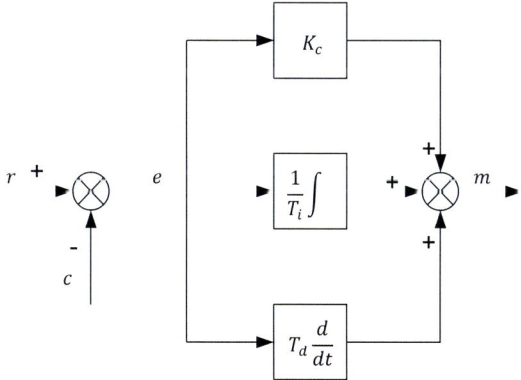

Figure 52. Parallel controller.

$$m(t) = K_c e(t) + \frac{1}{T_i} \int e(t)\, dt + T_d \frac{d}{dt} e(t) + b \qquad (2\text{-}12)$$

Exercise 2-1

Tuning and Control of a Pressure Loop

EXERCISE OBJECTIVE

Familiarize yourself with the use and manual tuning of P, PI, and on-off control schemes applied to pressure loops.

DISCUSSION OUTLINE

The Discussion of this exercise covers the following points:

- Recapitulation of relevant control schemes
- Tuning with the trial-and-error method

DISCUSSION

This exercise introduces three control schemes and puts them to use in a pressure process loop. This allows a comparative analysis of the different schemes in terms of efficiency, simplicity, and applicability to various situations. An intuitive method to tune controllers is also presented.

Recapitulation of relevant control schemes

A controller in proportional mode (P mode) outputs a signal ($m(t)$ – manipulated variable) which is proportional to the difference between the target value (SP: set point) and the actual value of the variable ($c(t)$ – controlled variable). This simple scheme works well but typically causes an offset. The only parameter to tune is the controller gain K_c (or the proportional band ($PB\% = 100\%/K_c$) if your controller uses this parameter instead).

A controller in proportional/integral mode (PI mode) works in a fashion similar to a controller in P mode, but also integrates the error over time to reduce the residual error to zero. The integral action tends to respond slowly to a change in error for large values of the integral time T_i and increases the risks of overshoot and instability for small values of T_i. Thus, the two parameters which require tuning for this control method are K_c (or $PB\%$) and T_i (or the integral gain, defined as $G_i = 1/T_i$).

The On-off control mode is the simplest control scheme available. It involves either a 0% or a 100% output signal from the controller based on the sign of the measured error. The option to add a dead band is available with most controllers to reduce the oscillation frequency and prevent premature wear of the final control element. There are no parameters to specify for this mode beyond a set point and dead band parameters. Note that it is possible to simulate an On-off mode with a controller in P mode for a large value of K_c (or a very small $PB\%$).

Tuning with the trial-and-error method

The **trial and error method** of controller tuning is a procedure to adjust the P, I, and D parameters until the controller is able to rapidly correct its output in response to a step change in the error. This correction is to be performed without excessive overshooting of the controlled variable.

This method is widely used because it does not require the characteristics of the process to be known and it is not required bringing the process into a sustained oscillation. Another important aspect of this method is that it is instrumental in developing an intuition for the effects of each of the tuning parameters.

However, the trial and error method can be daunting to perform for inexperienced technicians because a change in tuning constant tends to affect the action of all three controller terms. For example, increasing the integral action will increase the overshooting, which in turn will increase the rate of change of the error, which will then increase the derivative action. A structured approach and experience help in obtaining a good tuning relatively quickly without resorting to involved calculations.

A good trial-and-error method is to follow a geometrical progression in the search for optimal parameters. For example, multiplying or dividing one of the tuning parameters by two at each iteration can help you converge quickly toward an optimal value of the parameter.

A procedure for the trial-and-error method

The trial-and-error method is performed using the following procedure (also refer to Figure 53 and Figure 54 for PI control):

1. Set the controller in the mode you want to use: P, PI, PD, or PID. Follow the instructions to adjust every parameter relevant to the mode you are using. Note that you can use the PID mode to perform any of the modes by simply setting the parameters to appropriate values (e.g. $T_d = 0$ for PI mode).

Adjusting the P action

2. With the controller in manual mode, turn off the integral and derivative actions of the controller by setting T_i and T_d respectively to the largest possible value and 0.

3. Set the controller gain K_c to an arbitrary but small value, such as 1.

4. Place the controller in the automatic (closed-loop) mode.

5. Make a step change in the set point and observe the response of the controlled variable. The set point change should be typical of the expected use of the system.

   Since the controller gain is low, the controlled variable will take a relatively long time to stabilize (i.e. the response is likely to be overdamped).

6. Increase K_c by a factor of 2 and make another step change in the set point to see the effect on the response of the controlled variable.

> The controller gain K_c is related to the proportional band: $PB\% = 100\%/K_c$.
>
> If your controller uses the proportional band, start with a value of $PB\% = 100\%$ and replace instructions to increase K_c by a factor of two by a decrease of $PB\%$ by a factor of two.

Ex. 2-1 – Tuning and Control of a Pressure Loop ♦ *Discussion*

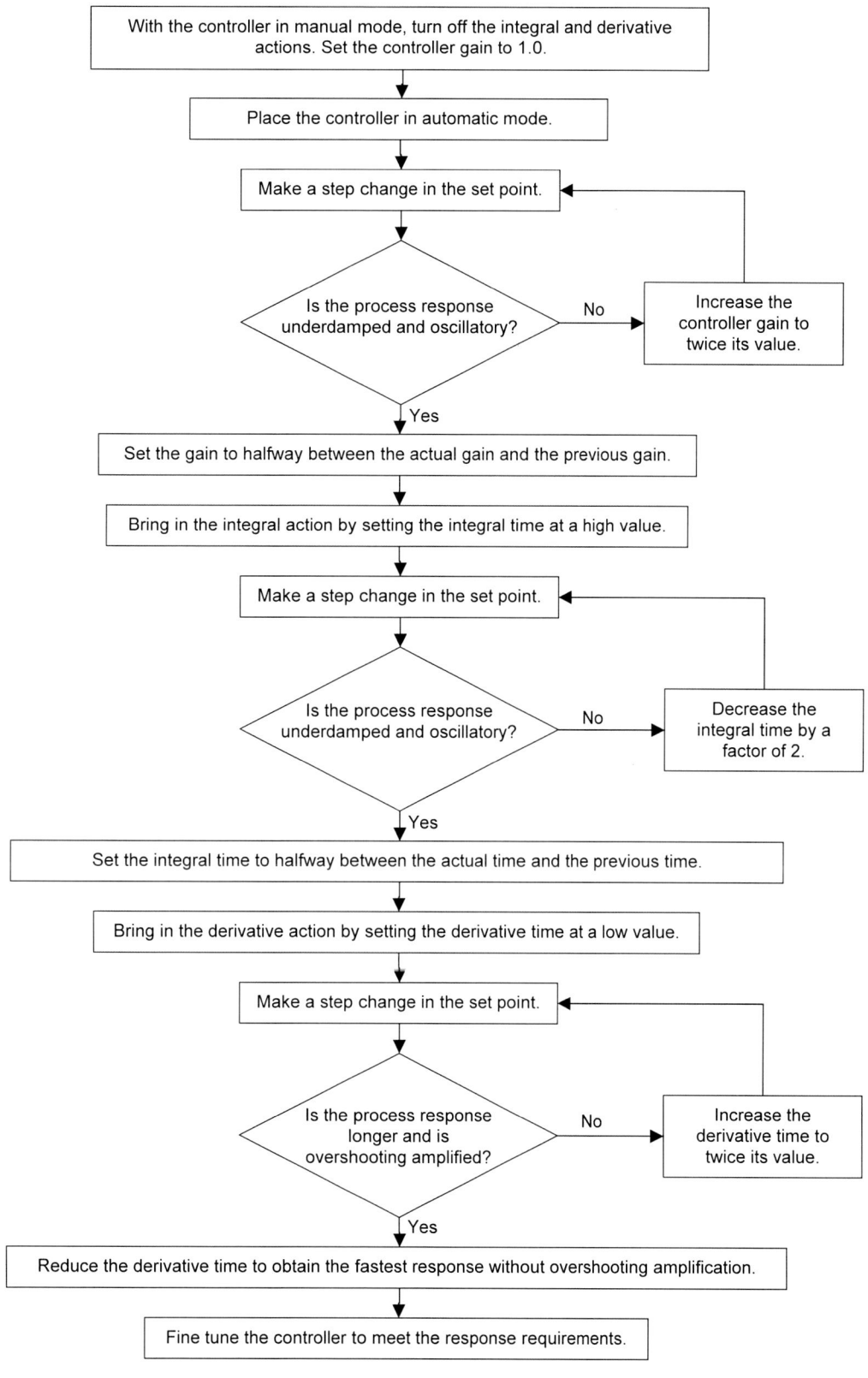

Figure 53. Trial-and-error tuning method.

The objective is to find the value of K_c at which the response becomes underdamped and oscillatory. This is the ultimate controller gain. Keep increasing K_c by factors of 2, performing a set point change after each new attempt, until you observe the oscillatory response.

Once the ultimate controller gain is reached, revert back to the previous value of K_c by decreasing the controller gain by a factor of 2. The P action is now set well enough to add another control action if required.

Adjusting the I action

7. Start bringing in integral action by setting the integral time T_i at an arbitrarily high value. Decrease T_i by factors of 2, making a set point change after each setting.

   Do so until you reach a value of T_i at which the response of the controlled variable becomes underdamped and oscillatory. At this point, revert back to the previous value of T_i by increasing T_i to twice its value.

   The I action is now set and you can now proceed to the adjustment of the D action if required.

Adjusting the D action

8. Start bringing in derivative action by setting the derivative time at an arbitrarily low value. Increase T_d by factors of 2, making a set point change after each setting.

   Do so until you reach the value of T_d that gives the fastest response without amplifying the overshooting or creating oscillation.

   The D action is now set.

Fine-tuning of the parameters

9. Fine-tune the controller until the requirements regarding the response time and overshooting of the controlled variable are satisfied.

A complementary approach to trial-and-error tuning

Another, more visual approach is to use Figure 54 to assist you in tuning your controller. The figure presents responses of a PI process to a step change for different combinations of parameters. A good tuning is shown in the center of the figure for 'optimal' K_c and T_i parameters. The tuning in the center is not necessarily the most appropriate for the process you want to control, but the response shown is a good target for a rough first tuning.

The figure also shows responses for detuned parameters (both above and below the 'optimal' K_c and T_i). Comparing the response you obtain for your system with the detuned responses in the figure tells you in which direction to change K_c, T_i, or both to converge towards the center case. Changing the parameters by a factor of two at every step until you get very close to the optimal value is a good method to converge rapidly.

Ex. 2-1 – Tuning and Control of a Pressure Loop ♦ *Procedure Outline*

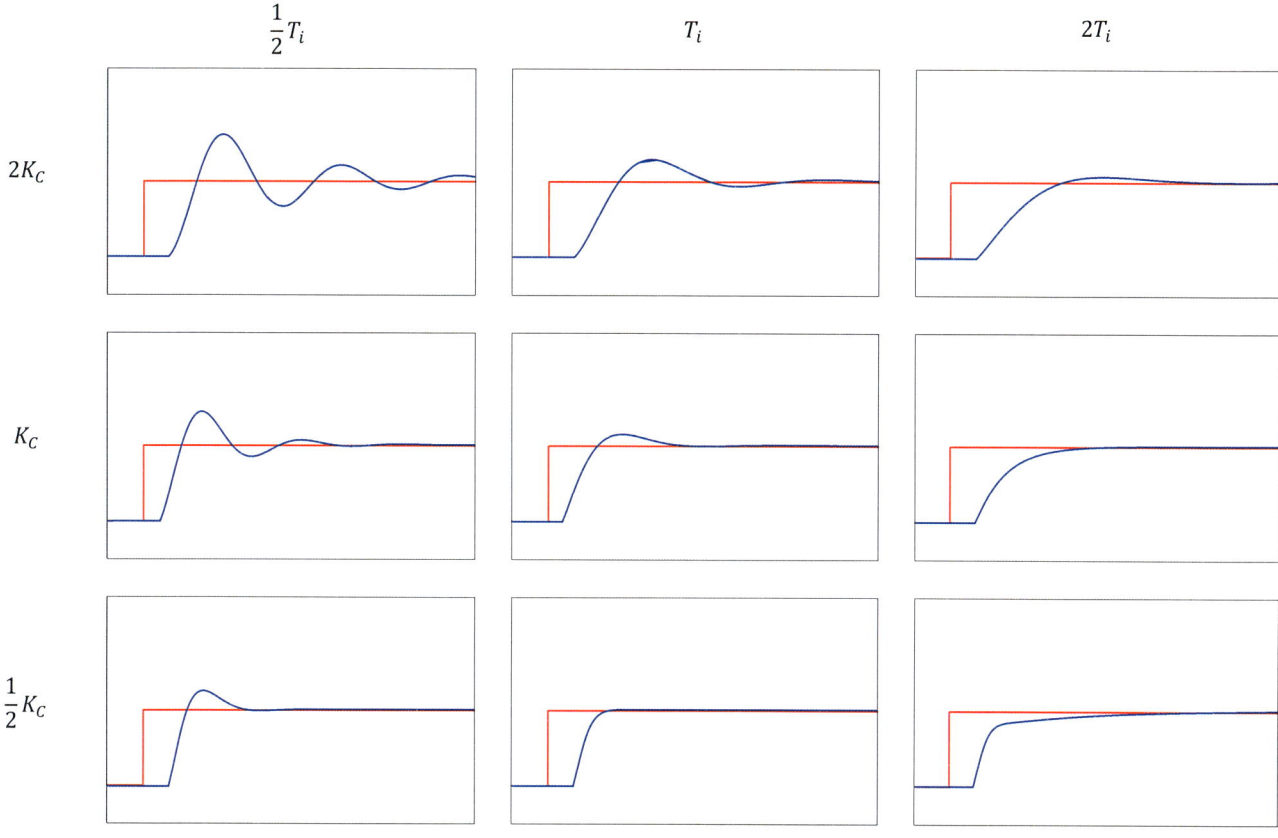

Figure 54. PID Tuning Chart.

Derivative action can then be added to the control scheme if required by following step 8 of the Trial-and-error method. Then, fine-tune the parameters to optimize the control and to meet the specific requirements of your process.

PROCEDURE OUTLINE   The Procedure is divided into the following sections:

- Set up and connections
- Adjusting the differential pressure transmitter
- Controlling the pressure loop
- Analyzing the results

PROCEDURE   **Set up and connections**

1. Connect the equipment according to the piping and instrumentation diagram (P&ID) shown in Figure 55 and use Figure 56 to position the equipment correctly on the frame of the training system.

Table 5. Material to add to the basic setup for this exercise.

Name	Model	Identification
Differential pressure transmitter (high-pressure range)	46920	PDIT 1
Solenoid valve	46951	S
Controller	*	PIC
Pressure control valve	46950-**	PCV

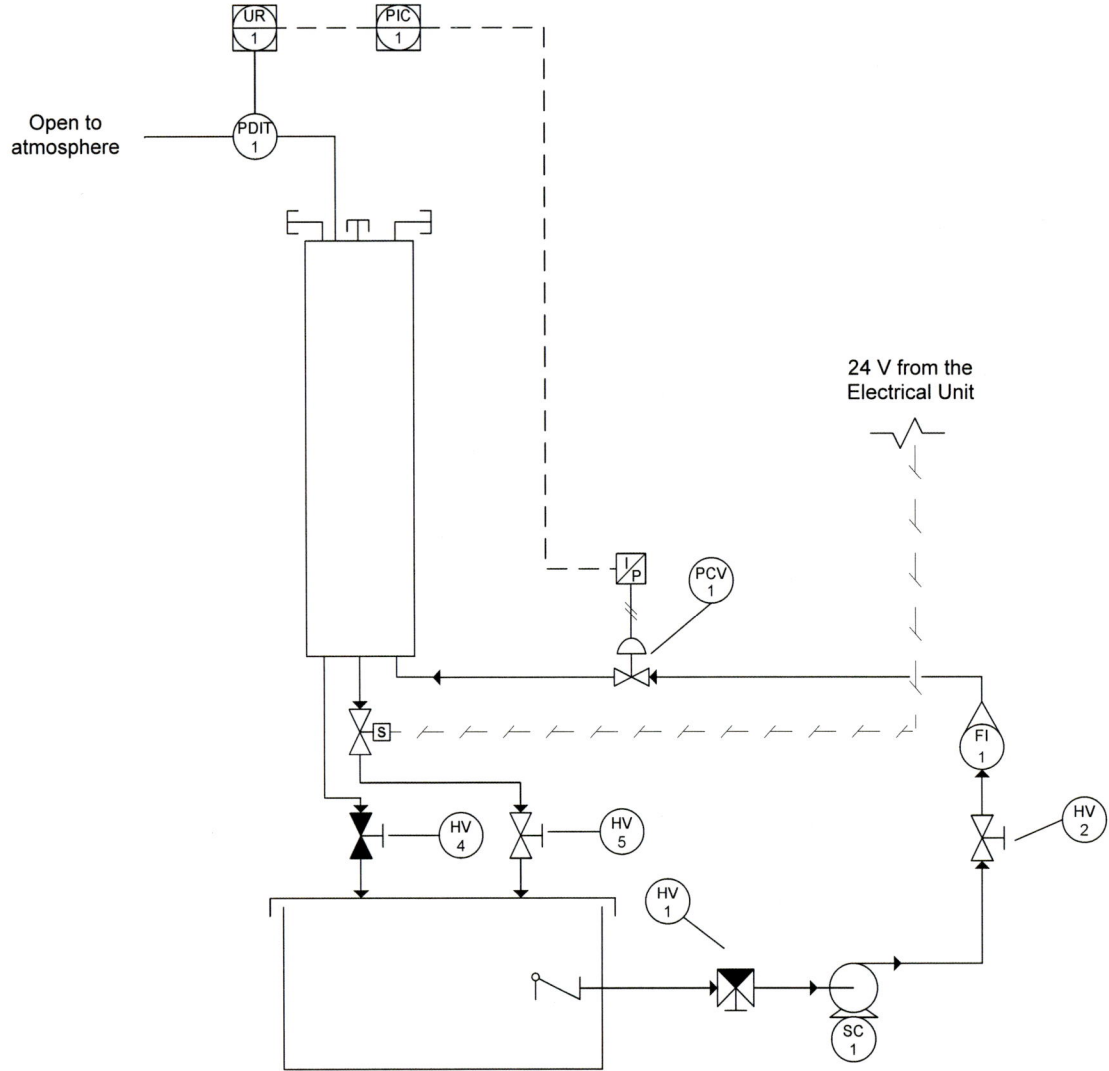

Figure 55. P&ID – Pressure control loop.

Ex. 2-1 – Tuning and Control of a Pressure Loop ♦ *Procedure*

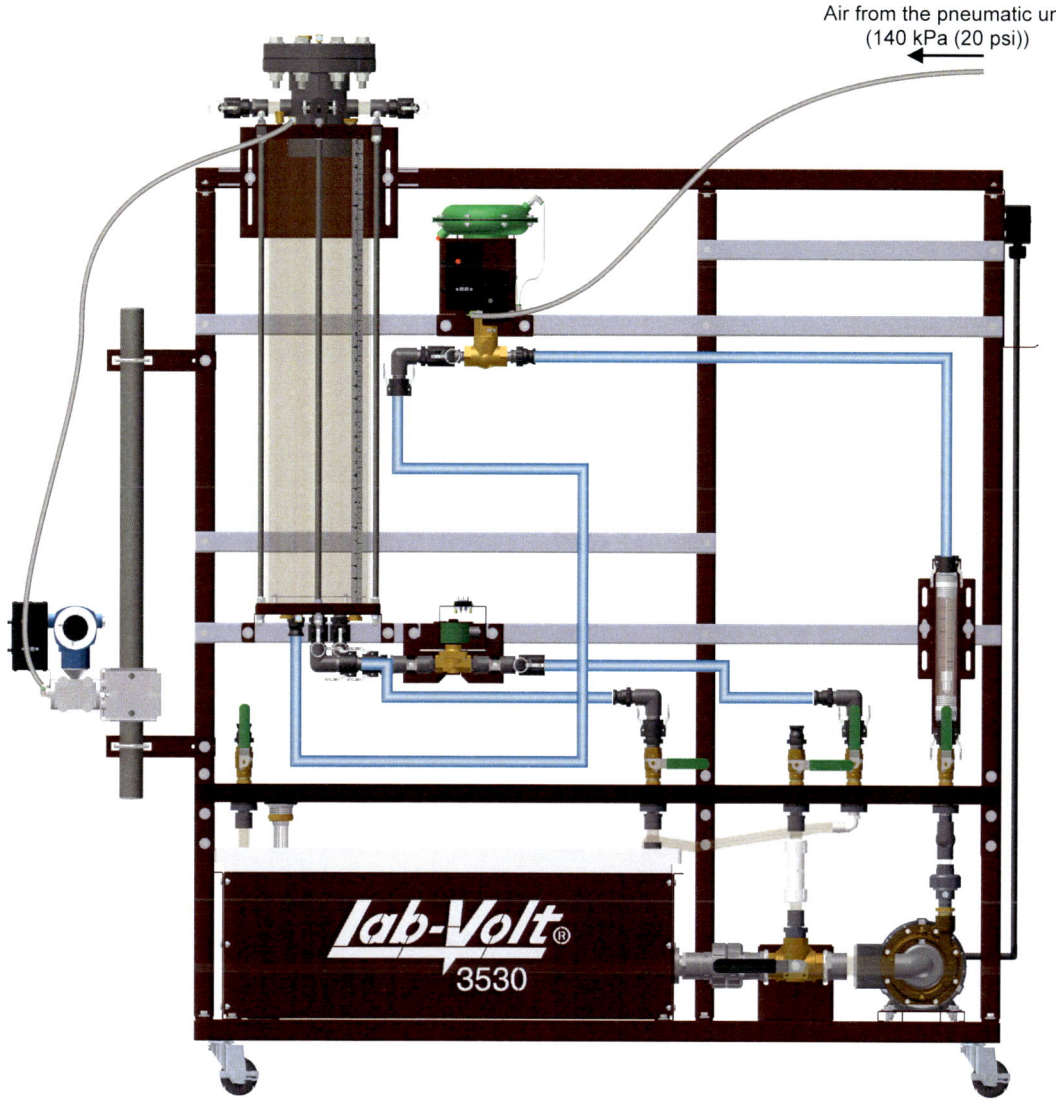

Figure 56. Setup – Pressure control loop.

2. Connect the control valve to the pneumatic unit.

3. Connect the pneumatic unit to a dry-air source with an output pressure of at least 700 kPa (100 psi).

4. Wire the emergency push-button so that you can cut power in case of emergency.

5. Do not power up the instrumentation workstation yet. You should not turn the electrical panel on before your instructor has validated your setup—that is not before step 12.

6. Connect the solenoid valve so that a voltage of 24 V dc actuates the solenoid when you turn the power on at step 12.

7. Connect the controller to the control valve and to the differential pressure transmitter. You must also include the recorder in your connections. On channel 1 of the recorder, plot the output signal from the controller and on channel 2, plot the signal from the transmitter. Be sure to use the analog input of your controller to connect the differential pressure transmitter.

8. Figure 57 shows how to connect the different devices together.

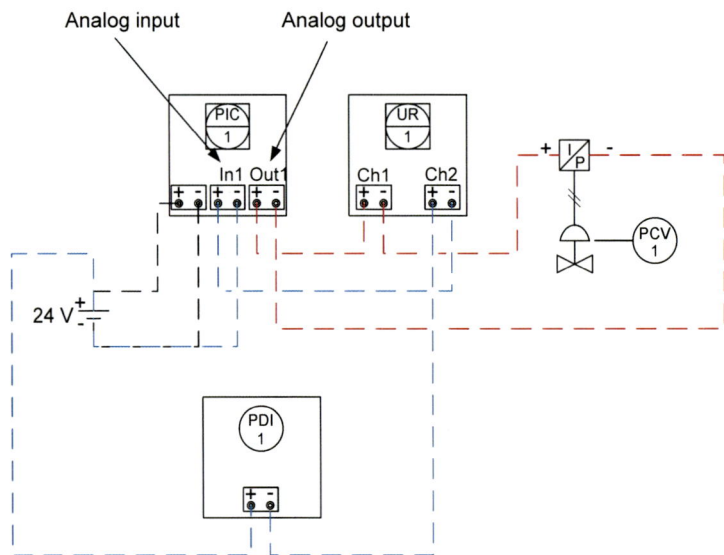

Figure 57. Connecting the equipment to the recorder.

9. Before proceeding further, complete the following checklist to make sure you have set up the system properly. The points on this checklist are crucial elements to the proper completion of this exercise. This checklist is not exhaustive, so be sure to follow the instructions in the *Familiarization with the Instrumentation and Process Control Training System* manual as well.

- All unused male adapters on the column are capped and the flange is properly tightened.
- The solenoid valve under the column is wired so that the valve opens when the system is turned on.
- The ball valves are in the positions shown in the P&ID.
- The three-way valve at the suction of the pump (HV1) is set so that the flow is directed toward the pump inlet.
- The control valve is fully open.

> ☑ The pneumatic connections are correct.
>
> ☑ The controller is properly connected to the differential pressure transmitter and to the control valve.
>
> ☑ The paperless recorder is connected correctly to plot the appropriate signals on channel 1 and channel 2.

10. Ask your instructor to check and approve your setup.

11. Remove one of the caps from the top of the column. This maintains the pressure in the column at the atmospheric pressure.

12. Power up the electrical unit, this starts all electrical devices as well as the pneumatic unit. Activate the control valve of the pneumatic unit to power the devices requiring compressed air.

13. In manual mode, set the output of the controller to 0%. The control valve should be fully open. If it is not, revise the electrical and pneumatic connections and make sure the calibration of the I/P converter is appropriate.

14. Test your system for leaks. Use the drive to make the pump run at low speed to produce a small flow rate. Gradually increase the flow rate, up to 50% of the maximum flow rate that the pumping unit can deliver (i.e., set the drive speed to 30 Hz). Repair any leaks and stop the pump.

Adjusting the differential pressure transmitter

Be sure to use the differential pressure transmitter, Model 46920. This differential pressure transmitter has a high-pressure range.

15. Make sure the impulse line of the differential pressure transmitter is free of water and that it is connected to the pressure port at the top of the column.

16. Configure the differential pressure transmitter so that it gives pressure readings in the desired units. Set transmitter parameters so that it sends a 4 mA signal if the pressure is 0 kPa (0 psi) and a 20 mA signal if the pressure is 32 kPa (4.6 psi).

17. Adjust the zero of the differential pressure transmitter. The column is at atmospheric pressure because of the removed cap; therefore the transmitter will read 0 kPa (0 psi) when the pressure inside the column is equal to the atmospheric pressure.

18. Replace the column cap removed at step 11. This will allow pressure to build in the column when you turn the pump on.

Controlling the pressure loop

19. Set the pump to 40.0 Hz and wait for the pressure reading to stabilize. Valve HV5 and the solenoid valve must be open.

P mode

20. Program the controller to operate in P mode. Tune the controller according to the trial-and-error method presented above. Note the value of K_c:

    $K_c =$

21. Record the response of the process to a step change in the set point of the controller from 40% to 60%. Transfer the data from the paperless recorder to a computer for later analysis.

PI mode

22. Program the controller to operate in PI mode. Tune the controller according to the trial-and-error method presented above. Note the value of K_c. and T_i:

    $K_c =$

    $T_i =$

23. Record the response of the process to a step change in the set point of the controller from 40% to 60%. Transfer the data from the paperless recorder to a computer for later analysis.

On-off mode

24. Program the controller to operate in On-off mode if such a mode is available with your controller. Experiment with different values of the dead band to visualize its effects. What do you observe as the dead band increases?

    _____

    _____

    Set the dead band to a value well suited to the process and which avoids excessive load on the control valve.

    If your controller does not have an On-off mode, simply set your controller in P mode with the largest possible K_c. The dead band typically cannot be adjusted in such cases.

25. Record the response of the process to a step change in the set point of the controller from 40% to 60%. Transfer the data from the paperless recorder to a computer for later analysis.

26. Stop the system.

Analyzing the results

27. Plot the response of the process for each mode using spreadsheet software. Compare the efficiency of the three modes and discuss their characteristics:

_____
_____
_____
_____
_____
_____
_____
_____

CONCLUSION

In this exercise, you learned to control a pressure loop using three different control modes: P, PI, and On-off. You experimented with the trial-and-error method of tuning a controller and developed a feel for the behavior of the control schemes for various values of the control parameters. The next exercise will cover a different method of optimizing a PID controller and will allow you to test your control skills on a flow process.

REVIEW QUESTIONS

1. What is the advantage of adding integral action to a proportional control scheme?

   _____
   _____
   _____

2. Why is On-off control not efficient in the experiment presented above?

   _____
   _____
   _____

3. Why does the trial-and-error method proceed with a factor of two change at every iteration?

4. What happens if you increase the K_c parameter in a PI control scheme?

5. What happens if you decrease the T_i parameter in a PI control scheme?

Exercise 2-2

Tuning and Control of a Flow Loop

EXERCISE OBJECTIVE

When you have completed this exercise, you will have gained experience with the use and tuning of PID control schemes applied to flow loops. The tuning method covered in this exercise is the ultimate period method.

DISCUSSION OUTLINE

The Discussion of this exercise covers the following points:

- Brief review of new control modes
- Tuning with the ultimate-cycle method
- Limits of the ultimate-cycle method

DISCUSSION

This exercise builds on the preceding one and introduces PID control schemes in the context of a flow process loop. The tuning of the controller is performed using the method of the **ultimate-cycle** (sometimes simply called **ultimate method**).

Brief review of new control modes

A controller in proportional, integral, and derivative mode (PID mode) incorporates the three control actions into a single polyvalent and powerful control scheme. The addition of derivative action to the PI mode covered in the previous exercise results in the capacity to attenuate overshoots to some extent, but adds the risk of instability if the process is noisy.

Tuning a controller in PID mode requires careful adjustment of the proportional gain (K_c), the integral time (T_i), and the derivative time (T_d) to properly address the control requirements of the process. In some circumstances the controller output must not be zero when the error is null, in theses cases a bias (b), also known as **manual reset** must be set.

The PD mode is also introduced in this experiment. It is similar to the P mode but adds the derivative action just described for the PID mode. It is not a widely used mode but is interesting to experiment with. Tuning a controller in PD mode implies adjusting the proportional gain (K_c) and the derivative time (T_d).

Tuning with the ultimate-cycle method

The ultimate-cycle tuning method is one of the first heuristic methods suggested by Ziegler and Nichols for tuning PID controllers (the method is consequently sometimes called the closed-loop Ziegler-Nichols method). The ultimate-cycle tuning method is designed to produce quarter-amplitude decay in the controlled variable after a given step change in the set point. This method enables the operator to calculate the P, I, and D tuning constants required for P, PI, PD, or PID control of a process using two parameters of the process: the ultimate gain (K_u), and the ultimate period (T_u).

The ultimate gain K_u is the largest value of K_c in P-only control mode such that the process is still stable (albeit marginally), i.e. the system is in a continuous, sustained oscillation. The **ultimate period** T_u is the period of the response when the gain is set to the ultimate gain.

> The ultimate proportional band PB_u can be used instead of K_u. It is then defined as the smallest value of PB for which the process is stable.
> $$PB\% = \frac{100\%}{K_c}$$

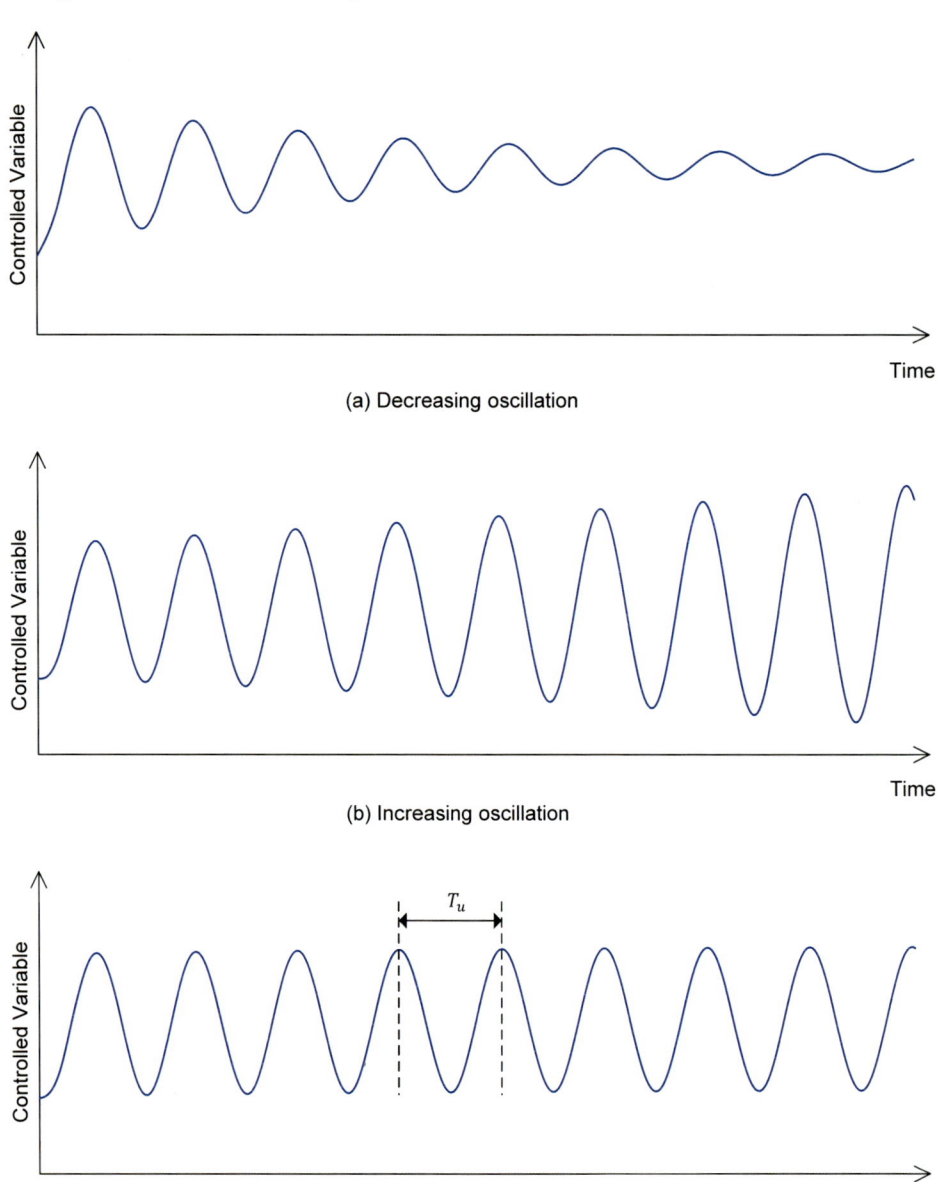

Figure 58. **Types of oscillations and determination of the ultimate period.**

The ultimate-cycle tuning method follows this procedure:

1. With the controller in manual mode, turn off the integral and derivative actions so as to use only P mode.

2. Set the proportional gain K_c at an arbitrary but somewhat small value, such as 1.

3. Place the controller in automatic (closed-loop) mode.

4. If the process starts to oscillate by itself, go to step 7. Otherwise, create a step change in the set point. The set point change should be typical of the expected use of the system.

5. If the process does not oscillate, increase the gain by a factor of 2.

6. Repeat steps 4 and 5 until the response becomes oscillatory.

7. Determine whether the oscillation is sustained—i.e. if it continues at the same amplitude without increasing or decreasing as in Figure 58 (c). If not, make small changes in the proportional gain until a sustained oscillation is achieved.

 Note: It is often necessary to wait for the completion of several oscillations before it can be determined if the oscillation is sustained.

The proportional gain, at which the sustained oscillation begins, without causing saturation of the controller output, is the ultimate proportional gain, K_u. Note this value. Then note the period of the oscillation of the process, as shown in Figure 58 (d). This is the ultimate period, T_u.

8. Using the ultimate proportional gain and ultimate period, calculate the tuning constants of the controller as follows:

Table 6. Control parameters for the ultimate-cycle tuning method.

Mode	Controller Gain K_c	Integral Time T_i	Derivative Time T_d
P	$K_c = 0.5K_u$ ($PB = 2PB_u$)	-	-
PI	$K_c = 0.45K_u$ ($PB = 2.2PB_u$)	$T_i = T_u/1.2$	-
PD	$K_c = 0.6K_u$ ($PB = 1.65PB_u$)	-	$T_d = T_u/8$
PID	$K_c = 0.6K_u$ ($PB = 1.65PB_u$)	$T_i = T_u/2.0$	$T_d = T_u/8$

Once the tuning constants of the controller are adjusted to the calculated values and the controller is returned in the automatic (closed-loop) mode, changes in the set point should produce a quarter-amplitude decay response. Optimization of the controller settings may require further fine-tuning.

Quarter-amplitude decay ratio

John G. Ziegler and Nathaniel B. Nichols, who were pioneers in control engineering, established a criterion to determine if a controller is appropriately tuned. This criterion is the **quarter-amplitude decay** ratio. Its states that, for two successive oscillations, the amplitude of the second oscillation should be one fourth of the amplitude of the first oscillation.

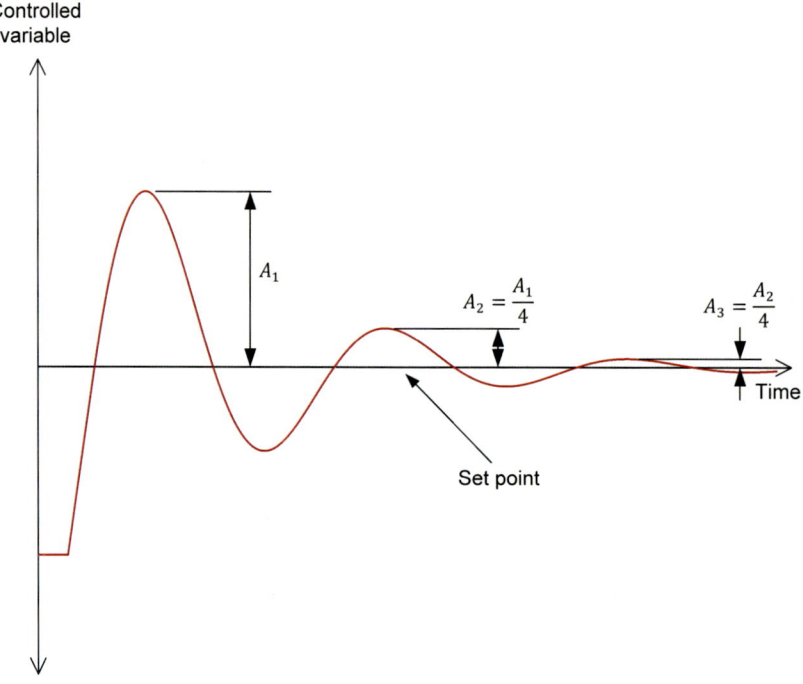

Figure 59. Quarter-amplitude decay ratio.

The quarter-amplitude decay response is a rough approximation for the optimal tuning of PID controllers. A controller is generally considered to be reasonably tuned when it satisfies this criterion, but fine tuning may be required to adapt the controller response to a specific process control application.

The quarter-amplitude decay response is a compromise between an underdamped and an overdamped response. The process response is **overdamped** when the controlled variable slowly returns to the set point after the step change without overshooting it. The response is **underdamped** when the controlled variable quickly returns to the set point with one or more overshoots before stabilizing. An underdamped response often means that the controller reacts too aggressively to correct the error, thereby overdoing it.

Limits of the ultimate-cycle method

It is important to note that the formulas given above apply only for non-interacting ideal controllers. Other formulas must be used for series or non-interacting parallel controllers. Refer to the section entitled Structure of controllers on page 47 for details.

It is also important to stress that using the ultimate-cycle tuning method may be out of the question in processes where bringing the system into continuous oscillation could be dangerous or might cause damage. Instead, another method of tuning, such as the trial-and-error method or the open-loop step response method, should be used. The open-loop step response method is also known as the open-loop Ziegler-Nichols method (this is covered in the next exercise).

Procedure Outline

The Procedure is divided into the following sections:

- Set up and connections
- Adjusting the differential pressure transmitter
- Controlling the flow loop

Procedure

Set up and connections

1. Connect the equipment according to the piping and instrumentation diagram (P&ID) shown in Figure 60 and use Figure 61 to position the equipment correctly on the frame of the training system.

Table 7. Material to add to the basic setup for this exercise.

Name	Model	Identification
Differential pressure transmitter (high-pressure range)	46920	PDIT 1
Solenoid valve	46951	S
Controller	*	FIC
Flow control valve	46950-**	FCV
Venturi tube	46911	FE 1
Three-valve manifold	85813	-

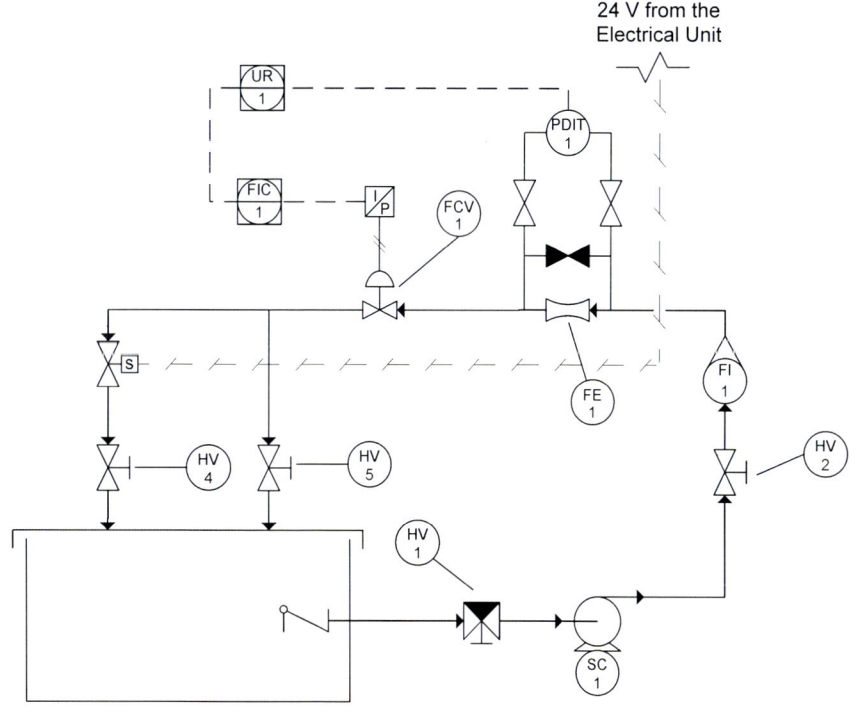

Figure 60. P&ID – Flow control loop.

Ex. 2-2 – Tuning and Control of a Flow Loop ♦ *Procedure*

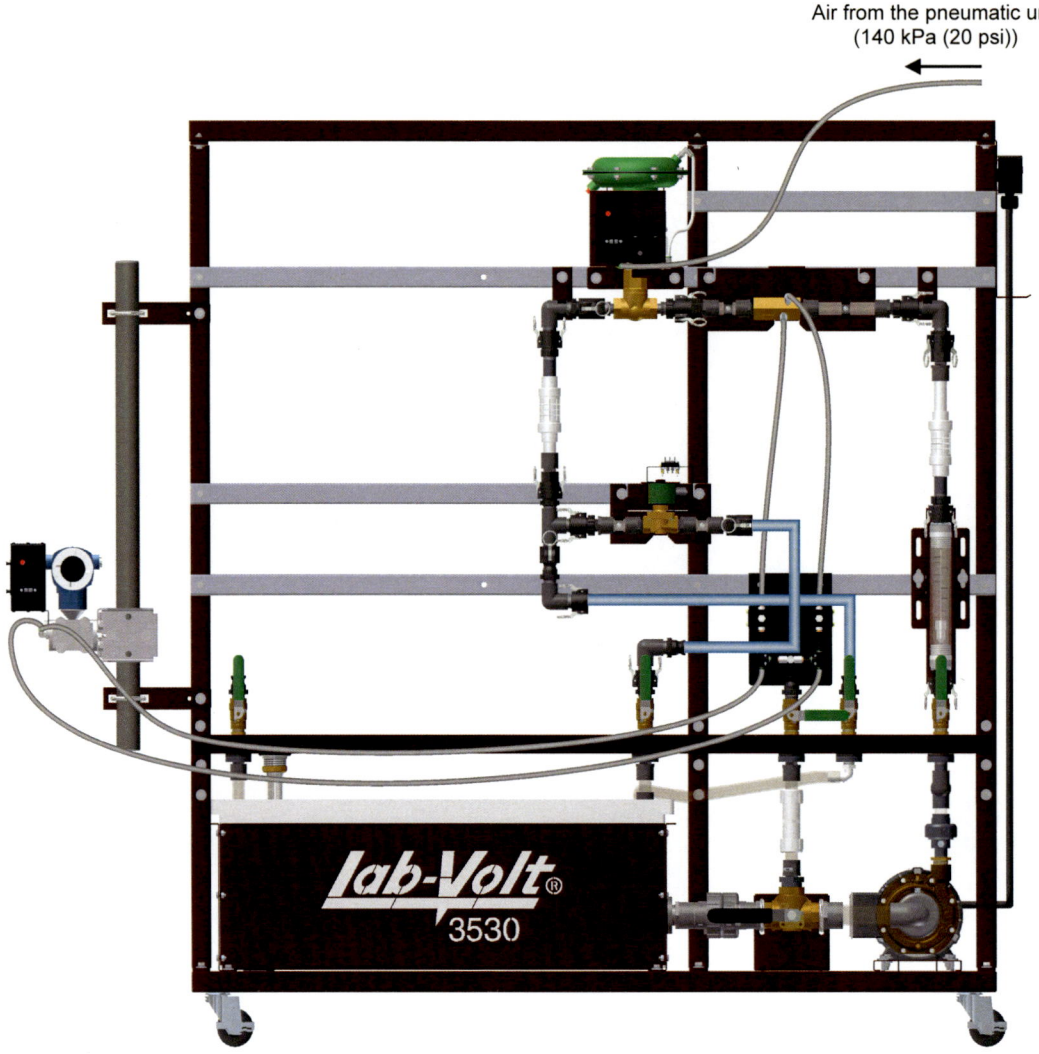

Figure 61. Setup – Flow control loop.

2. Connect the control valve to the pneumatic unit.

3. Connect the pneumatic unit to a dry-air source with an output pressure of at least 700 kPa (100 psi).

4. Wire the emergency push-button so that you can cut power in case of emergency.

5. Do not power up the instrumentation workstation yet. You should not turn the electrical panel on before your instructor has validated your setup—that is not before step 11.

6. Connect the solenoid valve so that a voltage of 24 V dc actuates the solenoid when you turn the power on at step 11.

7. Connect the controller to the control valve and to the differential pressure transmitter. You must also include the recorder in your connections. On channel 1 of the recorder, plot the output signal from the controller and on channel 2, plot the signal from the transmitter. Be sure to use the analog input of your controller to connect the differential pressure transmitter.

8. Figure 62 shows how to connect the different devices together.

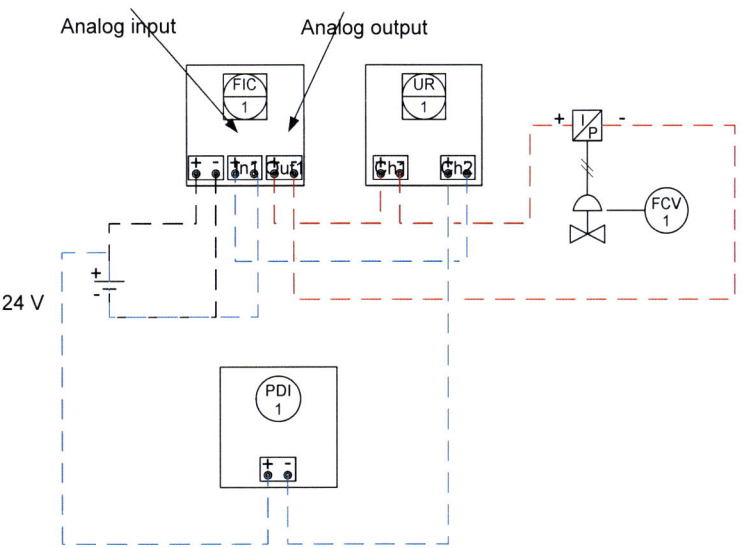

Figure 62. Connecting the equipment to the recorder.

9. Before proceeding further, complete the following checklist to make sure you have set up the system properly. The points on this checklist are crucial elements to the proper completion of this exercise. This checklist is not exhaustive, so be sure to follow the instructions in the *Familiarization with the Instrumentation and Process Control Training System* manual as well.

- The solenoid valve under the column is wired so that the valve opens when the system is turned on.
- The Venturi tube and the three-valve manifold are properly used according to the guidelines stated in the *Familiarization with the Instrumentation and Process Control Training System* manual.
- The ball valves are in the positions shown in the P&ID.
- The three-way valve at the suction side of the pump (HV1) is set so that the flow is directed toward the pump inlet.
- The control valve is fully open.
- The pneumatic connections are correct.

Ex. 2-2 – Tuning and Control of a Flow Loop ♦ *Procedure*

> 📷 The controller is properly connected to the differential pressure transmitter and to the control valve.
>
> 📷 The paperless recorder is connected correctly to plot the appropriate signals on channel 1 and channel 2.

10. Ask your instructor to check and approve your setup.

11. Power up the electrical unit, this starts all electrical devices as well as the pneumatic unit. Activate the control valve of the pneumatic unit to power the devices requiring compressed air.

12. In manual mode, set the output of the controller to 0%. The control valve should be fully open. If it is not, revise the electrical and pneumatic connections and make sure the calibration of the I/P converter is appropriate.

13. Test your system for leaks. Use the drive to make the pump run at low speed to produce a small flow rate. Gradually increase the flow rate, up to 50% of the maximum flow rate that the pumping unit can deliver (i.e., set the drive speed to 30 Hz). Repair any leaks and stop the pump.

Adjusting the differential pressure transmitter

Be sure to use the differential pressure transmitter, Model 46920. This differential pressure transmitter has a high-pressure range.

14. Be sure to connect the impulse lines of the differential pressure transmitter to the three-valve manifold. Bleed the impulse lines and configure the transmitter for flow measurement. Adjust the zero of the differential pressure transmitter.

    Set transmitter parameters so that a 4 mA signal is sent for a flow of 0 L/min (0 gal/min) and a 20 mA signal for a flow of 40 L/min (10 gal/min).

Controlling the flow loop

15. Set the pump to 40.0 Hz and wait for the flow reading to stabilize. Valves HV4, HV5, and the solenoid valve must be open.

16. Apply the ultimate-cycle method to determine the parameters K_u and T_u. The suggested step change in the set point should be from 40% to 60%. Accordingly, aim to find a sustained oscillation for a set point in the vicinity of 50%.

Note that it might be helpful and more precise to transfer the data to a computer in order to determine T_u using spreadsheet software.

$K_u =$

$T_u =$

17. Calculate the recommended parameters for your process and note the results in Table 8

Table 8. Calculated parameters for the ultimate-cycle tuning method.

Mode	Controller Gain K_c	Integral Time T_i	Derivative Time T_d
P	$K_c = 0.5K_u =$ $PB = 2PB_u =$	-	-
PI	$K_c = 0.45K_u =$ $PB = 2.2PB_u =$	$T_i = T_u/1.2 =$	-
PD	$K_c = 0.6K_u =$ $PB = 1.65PB_u =$	-	$T_d = T_u/8 =$
PID	$K_c = 0.6K_u =$ $PB = 1.65PB_u =$	$T_i = T_u/2.0 =$	$T_d = T_u/8 =$

Always make sure the units you use are in agreement with those of the controller. For instance, some controllers use units of minutes instead of seconds. Convert your results as required and develop the habit of checking for unit consistency whenever troubleshooting for unexpected behaviors.

18. For each control mode, program the controller according to the parameters you obtained in Table 8. Produce a step change in the set point from 40% to 60%. Sketch the resulting process responses and note your observations.

P mode

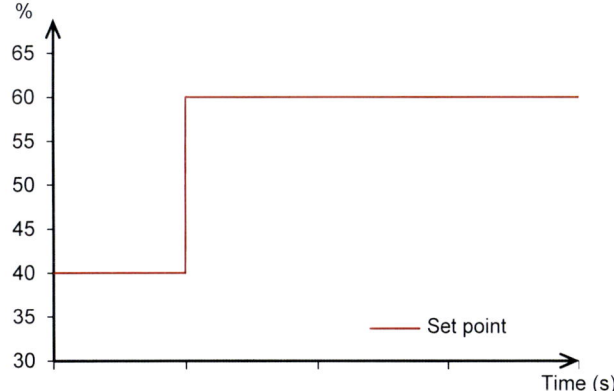

Figure 63. Sketch of the process response - P mode.

PI mode

Record the response in the case of the **PI mode** and transfer the resulting data to a computer.

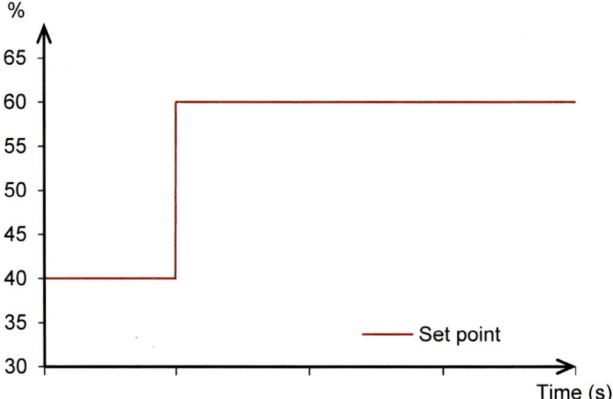

Figure 64. Sketch of the process response - PI mode.

PD mode

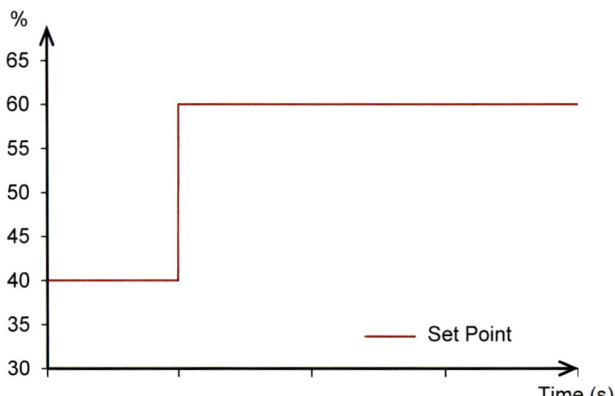

Figure 65. Sketch of the process response - PD mode.

PID mode

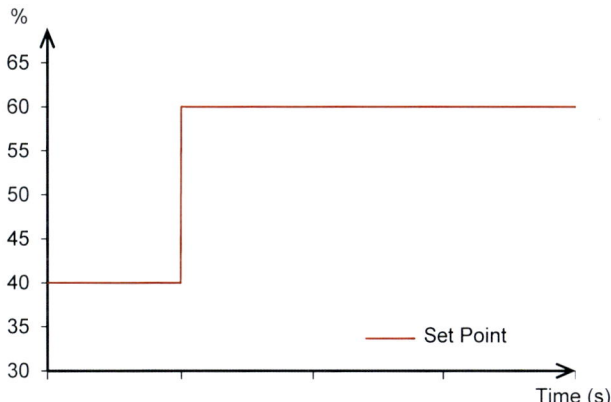

Figure 66. Sketch of the process response - PID mode.

Which of the four modes would you choose to control this process? Why?

19. Imagine your process loop controls the flow of coolant in an industrial cryogenics process. It is critical to avoid oscillations of the process variable about the set point as this would result in thermal shocks to the system. Try fine-tuning the parameters of your process in the PI mode to avoid oscillations while retaining as fast a response as possible.

    Can you modify the response of the process adequately? If so, record and transfer your improved response in PI mode to a computer. Plot graphs of both responses in PI mode (the standard one obtained previously and your fine-tuned version).

    How practical would you say the ultimate-cycle tuning method is for this particular process?

20. Experiment with the PID mode for different values of the T_d parameter. Use the values obtained with the ultimate-cycle method for both K_c and T_i. What do you observe as you keep increasing T_d?

21. Stop the system, turn off the power, and store the equipment as required.

Ex. 2-2 – Tuning and Control of a Flow Loop ♦ *Conclusion*

CONCLUSION

In this exercise, you learned to control a flow loop using different control modes: P, PI, PD, and PID. You learned to use the ultimate-cycle method of tuning a controller and developed a feel for the behavior of the control schemes for various values of the control parameters. The next exercise will cover a different method to optimize a PID controller and will allow you to test your control skills with a level process.

REVIEW QUESTIONS

1. What are the required parameters for the ultimate-cycle method?

2. How does one find the value of those parameters?

3. What does 'sustained oscillation' mean?

4. When would it be unsuitable to tune a process via the ultimate-cycle method?

5. What happens if you increase T_d too much in PID control mode?

Exercise 2-3

Tuning and Control of a Level Loop

EXERCISE OBJECTIVE

When you have completed this exercise, you will have gained experience with the use and tuning of PID control schemes applied to level loops. The tuning method covered in this exercise is the open-loop Ziegler-Nichols method.

DISCUSSION OUTLINE

The Discussion of this exercise covers the following point:

- The open-loop Ziegler-Nichols method

DISCUSSION

This exercise presents one more method of tuning a controller. This method is based on the characteristics of the process reaction which are readily obtained from the response to a step change. The exercise then applies the open-loop Ziegler-Nichols method to a level process.

The open-loop Ziegler-Nichols method

This method of controller tuning was developed in 1942 by John G. Ziegler and Nathaniel B. Nichols. It enables the operator to calculate the P, I, and D tuning constants required for P, PI, or PID control of a process based on the open-loop response of the process to a step change in the set point.

The open-loop step response method is performed according to the following procedure:

1. With the controller in open-loop mode, create a step change in controller output. The resulting change in controlled variable should be typical of the expected use of the system. Note that you can use a calibrator instead of the controller to create a step change.

2. Based on the response curve of the controlled variable, determine the process gain K_p, the dead time t_d, and the time constant τ of the process. Refer to Ex. 1-1 for a discussion on process parameters.

Calculate the value of the parameter κ

$$\kappa = \left| \frac{\tau}{t_d K_p} \right|$$

3. Using the process characteristics found in step 2, calculate the tuning constants of the controller as follows:

Table 9. Control parameters for the open-loop Ziegler-Nichols tuning method.

Mode	Proportional Gain K_c	Integral Time T_i	Derivative Time T_d
P	$K_c = \kappa$	-	-
PI	$K_c = 0.9\,\kappa$	$T_i = 3.33\,t_d$	-
PID	$K_c = 1.2\,\kappa$	$T_i = 2\,t_d$	$T_d = 0.5\,t_d$

Once the tuning constants of the controller are adjusted to the calculated values and the controller is returned to the closed-loop mode, a typical change in the set point should produce the desired quarter-amplitude decay response. The controller should also be able to correct for load changes rapidly without excessive overshooting or oscillation of the controlled variable. Note, however, that small readjustments of the P, I, and D tuning constants may be required to obtain the optimum controller setting.

It is important to note that the formulas given above apply only to non-interacting, ideal controllers. Other formulas must be used for series or non-interacting parallel controllers. Refer to the section entitled Structure of controllers on page 47 for details.

An advantage of the open-loop step response method is that the process needs to be disturbed only once to obtain the required process characteristics. On the other hand, the determination of precise process parameters is a bit more involved and requires a few calculations.

PROCEDURE OUTLINE

The Procedure is divided into the following sections:

- Set up and connections
- Adjusting the differential pressure transmitter
- Applying the open-loop Ziegler-Nichols tuning method
- Controlling the level loop

PROCEDURE

Set up and connections

1. Connect the equipment according to the piping and instrumentation diagram (P&ID) shown in Figure 67 and use Figure 68 to position the equipment correctly on the frame of the training system.

Table 10. Material to add to the basic setup for this exercise.

Name	Model	Identification
Differential pressure transmitter (high-pressure range)	46920	LIT 1
Solenoid valve	46951	S
Controller	*	LIC
Level control valve	46950-**	LCV

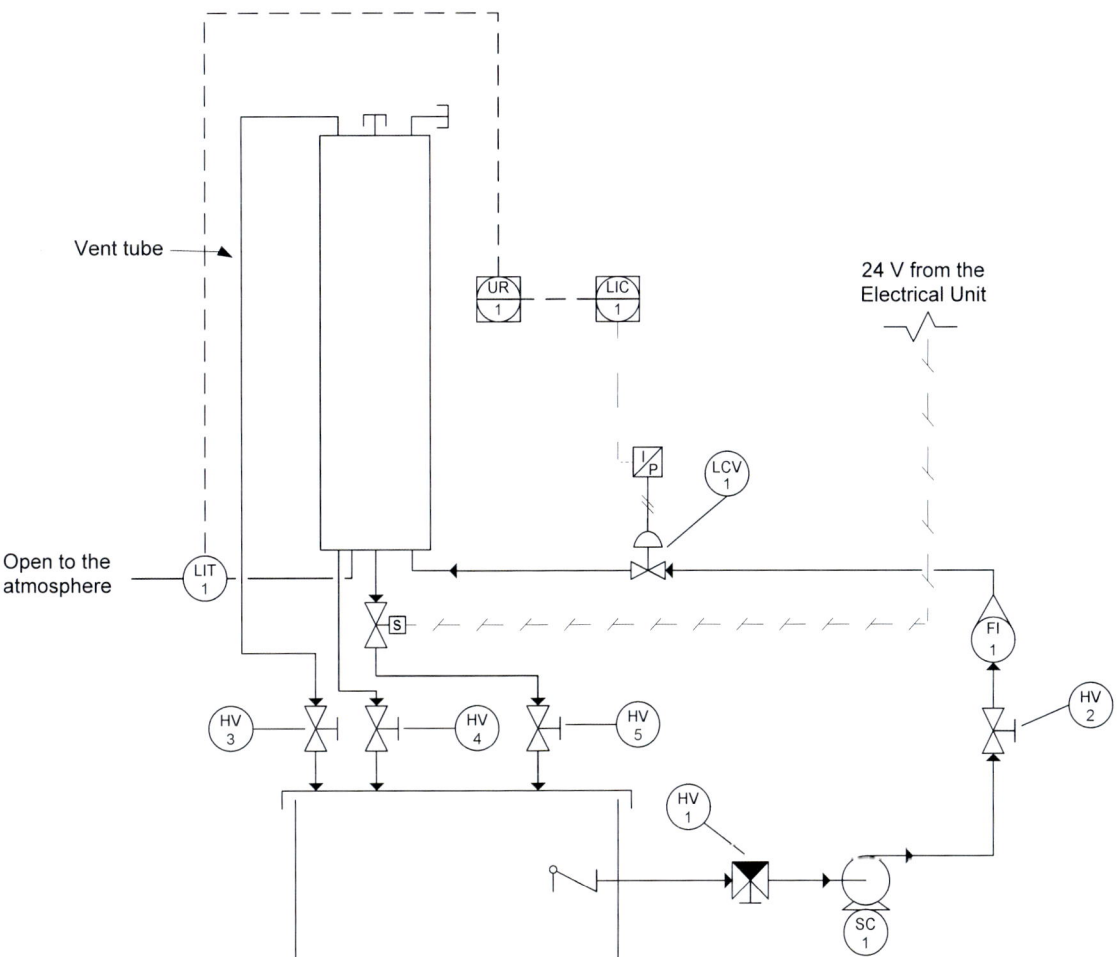

Figure 67. P&ID – Level control loop.

Ex. 2-3 – Tuning and Control of a Level Loop ♦ *Procedure*

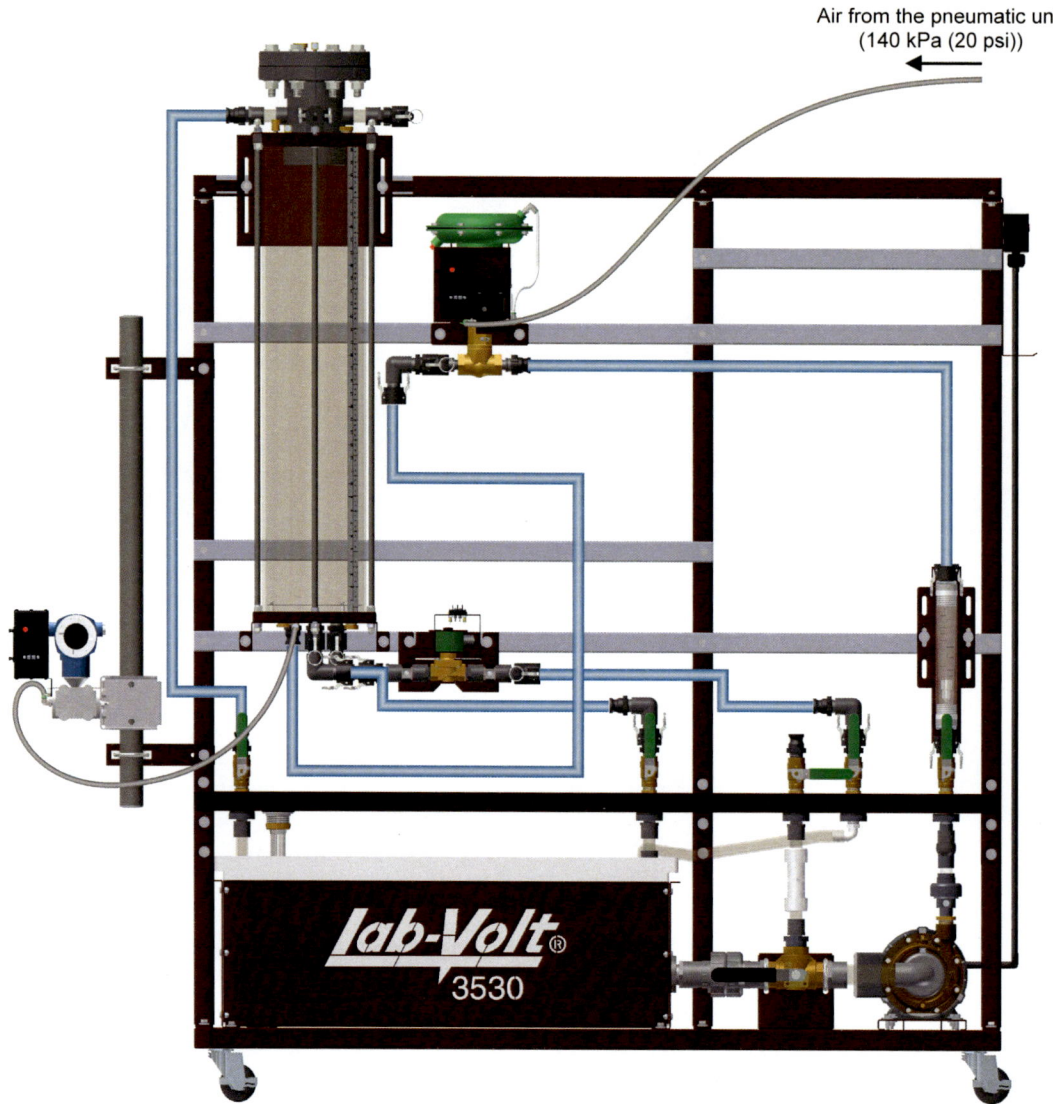

Figure 68. Setup – Level control loop.

2. Connect the control valve to the pneumatic unit.

3. Connect the pneumatic unit to a dry-air source with an output pressure of at least 700 kPa (100 psi).

4. Wire the emergency push-button so that you can cut power in case of emergency.

5. Do not power up the instrumentation workstation yet. You should not turn the electrical panel on before your instructor has validated your setup—that is not before step 11.

Ex. 2-3 – Tuning and Control of a Level Loop ♦ *Procedure*

6. Connect the solenoid valve so that a voltage of 24 V dc actuates the solenoid when you turn the power on at step 11.

7. Connect the controller to the control valve and to the differential pressure transmitter. You must also include the recorder in your connections. On channel 1 of the recorder, plot the output signal from the controller and on channel 2, plot the signal from the transmitter. Be sure to use the analog input of your controller to connect the differential pressure transmitter.

8. Figure 69 shows how to connect the different devices together.

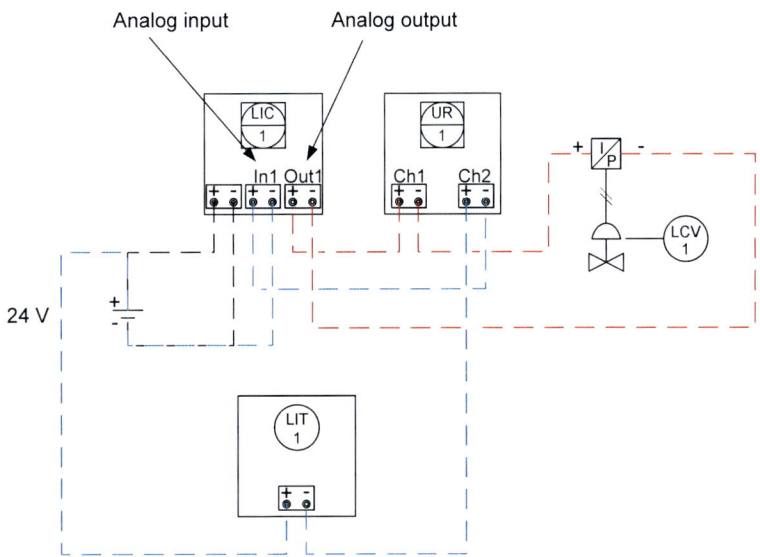

Figure 69. **Connecting the equipment to the recorder.**

9. Before proceeding further, complete the following checklist to make sure you have set up the system properly. The points on this checklist are crucial elements to the proper completion of this exercise. This checklist is not exhaustive, so be sure to follow the instructions in the *Familiarization with the Instrumentation and Process Control Training System* manual as well.

- The solenoid valve under the column is wired so that the valve opens when the system is turned on.
- All unused ports on the column are capped.
- The ball valves are in the positions shown in the P&ID.
- The three-way valve at the suction side of the pump (HV1) is set so that the flow is directed toward the pump inlet.
- The control valve is fully open.
- The pneumatic connections are correct.
- The controller is properly connected to the differential pressure transmitter and to the control valve.

> 📼 The paperless recorder is connected correctly to plot the appropriate signals on channel 1 and channel 2.

10. Ask your instructor to check and approve your setup.

11. Power up the electrical unit, this starts all electrical devices as well as the pneumatic unit. Activate the control valve of the pneumatic unit to power the devices requiring compressed air.

12. In manual mode, set the output of the controller to 0% and 100%. The control valve should be fully open in the first case and fully closed in the second case. If it is not, revise the electrical and pneumatic connections and make sure the calibration of the I/P converter is appropriate.

13. Test your system for leaks. Use the drive to make the pump run at low speed to produce a small flow rate. Gradually increase the flow rate, up to 50% of the maximum flow rate that the pumping unit can deliver (i.e., set the drive speed to 30 Hz). Stop the pump and repair any leaks.

Adjusting the differential pressure transmitter

14. Bleed the impulse line and configure the transmitter for level measurement. Adjust the zero of the differential pressure transmitter.

    Set transmitter parameters so that a 4 mA signal is sent for a level of 0 m (0 in) and a 20 mA signal for a level of 0.75 m (30 in).

Applying the open-loop Ziegler-Nichols tuning method

15. In manual mode, set the controller to output a signal of 40%.

16. Set the pump to 50.0 Hz. Valves HV3, HV4, and HV5, as well as the solenoid valve, must be open.

17. Let the process stabilize (be patient!) and then make an output step change to 50%. Let the process stabilize again and transfer the data recorded by the paperless recorder to a computer. The whole process takes about 15 minutes.

     Note that you can use a calibrator in lieu of the controller to perform the output step change.

18. Use your favorite method from Ex. 1-1 to determine the process parameters and note them here:

$t_d =$ $\qquad$ $\tau =$ $\qquad$ $K_p =$

Use these parameters to calculate κ:

$$\kappa = \left| \frac{\tau}{t_d K_p} \right| =$$

19. From the process characteristics you just obtained, you can now calculate the PID coefficient prescribed by the open-loop Ziegler-Nichols method. Note your results in Table 11.

Table 11. Calculated control parameters for the Ziegler-Nichols method.

Mode	Proportional Gain K_c	Integral Time T_i	Derivative Time T_d
P	$K_c = \kappa =$	-	-
PI	$K_c = 0.9\,\kappa =$	$T_i = 3.33\,t_d =$	-
PID	$K_c = 1.2\,\kappa =$	$T_i = 2\,t_d =$	$T_d = 0.5\,t_d =$

Be careful with units. Your controller might use units different from those you used in Table 11.

Controlling the level loop

20. Set the controller to operate in PID mode with the parameters you just calculated.

    Create a 40% to 60% step change in the set point and observe the evolution of the system. Record and transfer the data to a computer. Plot a graph of your results.

21. Try different set-point values by increasing or decreasing the set point by step changes of 20%. Determine whether the controller tuning remains acceptable over a broad range of set points. Fine-tune the parameters if necessary.

22. Once the system is well tuned, set the controller to a set point of 50%.

Create a sudden change in process load by closing valve HV4. Is the controller able to rapidly correct for the load change without oscillation of the controlled variable?

_____

_____

How much time is required for the process to return to the set point?

_____

23. Re-open valve HV4 and let the process stabilize again to the set point of 50%. Suddenly increase the drive speed to 60 Hz to create a disturbance in the flow input.

Is the system reacting in the same way as in the previous step? Can it handle the load change? How much time is required for the process to return to the set point?

_____

_____

_____

24. Stop the pump and empty the column.

25. Stop the system, turn off the power, and store the equipment.

CONCLUSION

This experiment presented the control of a level process loop, the third and last basic type of control covered in this manual. You also became acquainted with the widely used open-loop Ziegler-Nichols tuning method which completes your tool kit of basic tuning methods. The next exercise will explore a more advanced control scheme: Cascade control.

REVIEW QUESTIONS

1. Briefly describe the procedure to follow in applying the open-loop Ziegler-Nichols tuning method.

_____

_____

_____

_____

2. Which process characteristic determines both the integral time and the derivative time when using the open-loop Ziegler-Nichols tuning method?

_____

3. Cite an advantage and a disadvantage of the tuning method presented in this exercise.

4. How is the vent tube used in this setup?

5. Would the process gain (K_p) and the time constant (τ) be affected by the use of a column of larger diameter?

Exercise 2-4

Cascade Control of a Level/Flow Process

EXERCISE OBJECTIVE

When you have completed this exercise, you will be familiar with the concept of cascade control and its application to a level/flow process.

DISCUSSION OUTLINE

The Discussion of this exercise covers the following points:

- Cascade control
- Tuning a cascade control system

DISCUSSION

The PID control systems we have studied thus far have used a single control loop. Although single-loop control may provide satisfactory results, other forms of advanced control are available and may prove to be more advantageous. Among these is **cascade control**.

Cascade control

The different measureable variables in a process are often interrelated to the extent that knowledge of an intermediate variable could be used beneficially to control a primary process variable. One example is a flow of water entering a tank in which the level is to be controlled. Controlling the incoming flow with a control valve certainly has a direct effect on the level of water in the tank. This is because the two variables, incoming flow and tank level, are interacting. Cascade control takes advantage of this inherent interaction between variables to improve the overall control on a process variable.

Cascade control utilizes two control loops: a master loop and a slave loop. The **master loop** contains the primary, or **master controller** and monitors the primary variable. The **slave loop** contains a secondary, or **slave controller** which monitors a second variable. The output of the master controller is connected to the set-point input of the slave controller, causing the two controllers to be cascaded.

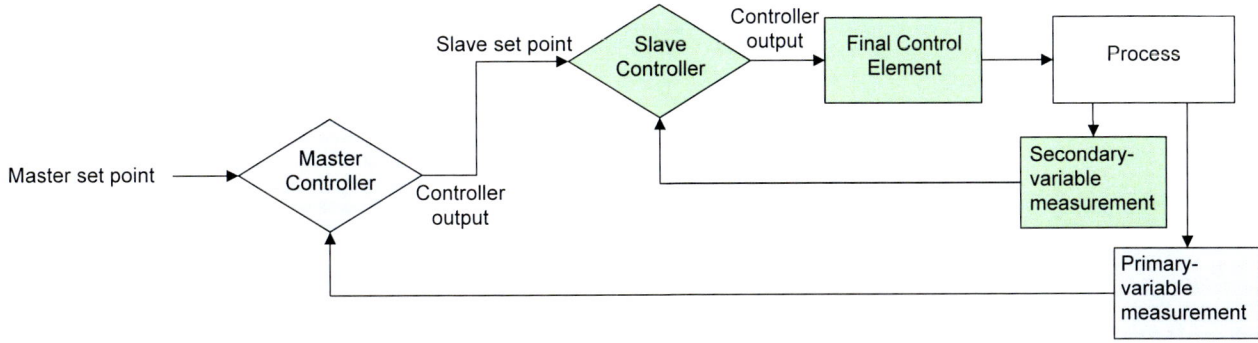

Figure 70. Block diagram of a general cascade-control scheme.

The main purpose of cascade control is to minimize the disturbances that affect the secondary variable before they cause pronounced changes in the primary controlled variable. Another advantage is the improvement in the speed of response of the secondary variable.

The principle of cascade control is best illustrated by an example involving the control of the level of fluid in a tank through regulation of the output flow rate (the input flow rate is assumed to be constant). The typical single-loop control scheme is illustrated in Figure 71. In this case, the level of water in the tank is monitored directly by a single controller which activates the control valve to steer the measured level towards the set-point value. A sudden change in the output flow due to a change in the downstream load would result in a level change. The controller however cannot respond to this new situation until the new flow rate has had a measurable effect on the level in the tank.

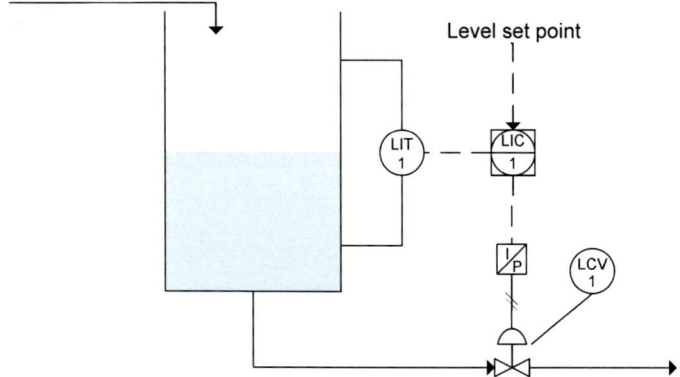

Figure 71. Simple level control.

Figure 72 shows how the cascade scheme works around this shortcoming by using a second loop with a slave controller which monitors and controls the flow at the output of the system. The master controller still monitors the level in the tank but adjusts the set point of the slave controller instead of the control valve. The flow control loop regulates load changes before they have an important impact on the level.

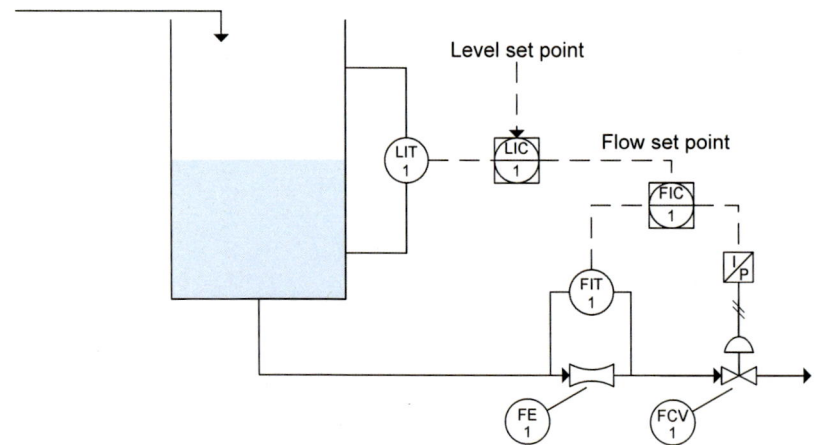

Figure 72. Cascade control (Level/Flow).

Thus, a cascade control system does not have to wait for the primary controlled variable to change before initiating corrective action. A change in the secondary controlled variable is sufficient to do so. Although a variation of the primary controlled variable can occur, this variation is typically not as important when it is under cascade control.

Cascade control is effective when the slave loop is more responsive than the master loop. A general rule of thumb states that the *slave-loop time constant should be smaller than four to ten times the time constant of the master loop*. If this is not the case, cascade control should be avoided because the system will tend to be unstable. Notice that it is also possible to cascade multiple loops in series as long as relevant process variables can be measured.

Tuning a cascade control system

The tuning of a cascade control system can be done as follows:

1. With the master controller in manual mode, the slave controller is tuned first. Normally, proportional (P) action only is utilized for the slave controller. However, integral (I) action is sometimes used when the process has a short time constant, as in the case of flow processes.

   Since the slave loop can be treated as if it were a single-control loop, the P and I (if any) constants of the slave controller can be determined using any of the previously presented methods of controller tuning.

2. Once the slave controller has been tuned, it is switched into automatic mode and the master controller is set for PI or PID control. The master loop can be considered a single-control loop because the slave loop can be treated as a final control element. Based on this assumption, the master controller can be tuned using any of the methods previously presented.

Secondary control modes

A few suggestions on the different control modes appropriate for the secondary loop (slave loop) in different specific situations:

- Control of flow is normally done in PI mode

- Valve positioners are usually controlled in proportional mode with a large gain ($K_c \approx 20$).

- Derivative action should be avoided if it acts on set-point changes as this causes overshoot. Some controllers are capable of applying the derivative action on the controlled variable only. Such a controller can be used profitably with derivative action in a secondary loop when the measured variable is not too tainted by noise.

PROCEDURE OUTLINE The Procedure is divided into the following sections:

- Set up and connections
- Adjusting the differential pressure transmitters
- Tuning the slave loop
- Tuning the master loop
- Controlling the level/flow cascade loop

PROCEDURE

Set up and connections

1. Connect the equipment according to the piping and instrumentation diagram (P&ID) shown in Figure 73 and use Figure 74 to position the equipment correctly on the frame of the training system.

Table 12. Material to add to the basic setup for this exercise.

Name	Model	Identification
Differential pressure transmitter (high-pressure range)	46920	FIT 1
Differential pressure transmitter (low-pressure range)	46921	LIT 1
Solenoid valve	46951	S
Controller (2 loops in cascade)	*	LIC
Flow control valve	46950-**	FCV
Venturi tube	46911	FE 1
Three-valve manifold	85813	-

Ex. 2-4 – Cascade Control of a Level/Flow Process ♦ *Procedure*

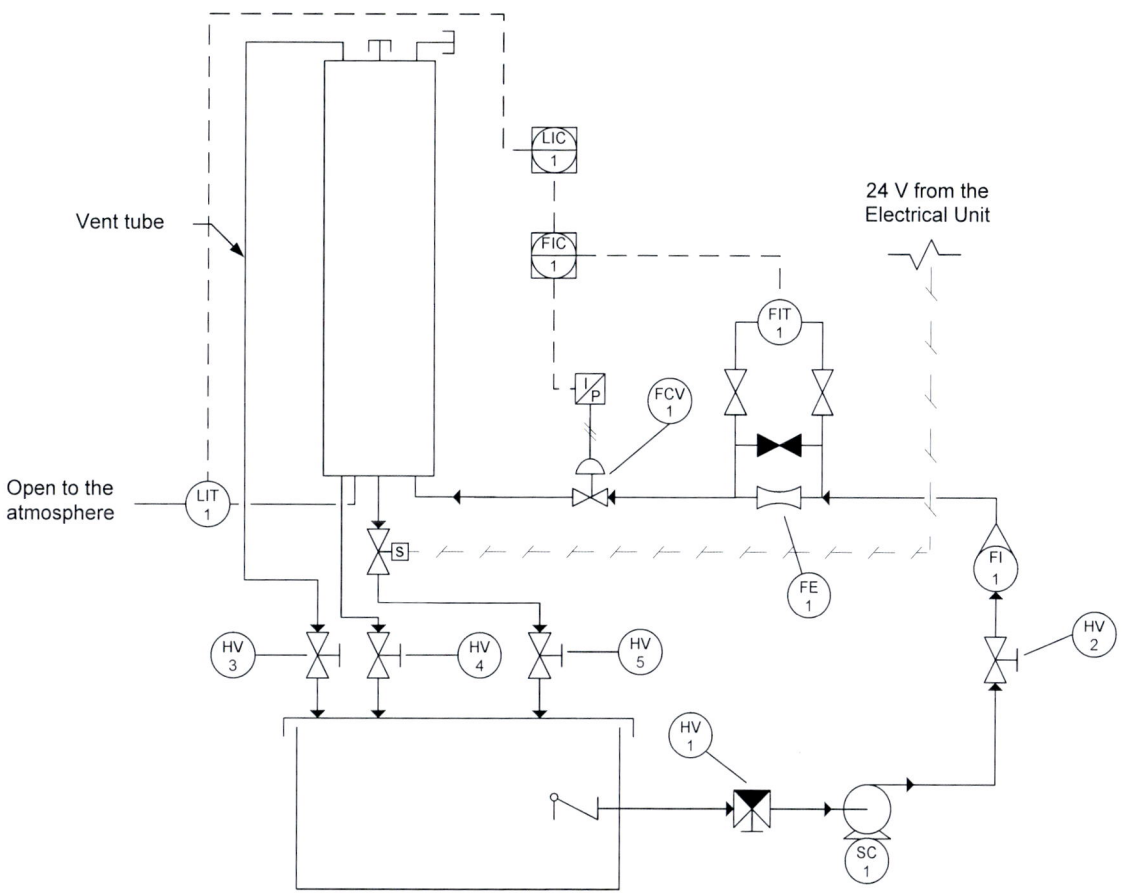

Figure 73. P&ID – Cascade control loop.

 The paperless recorder (UR) is not displayed in the P&ID above. See Figure 75 for the suggested electrical connections.

Ex. 2-4 – Cascade Control of a Level/Flow Process ♦ *Procedure*

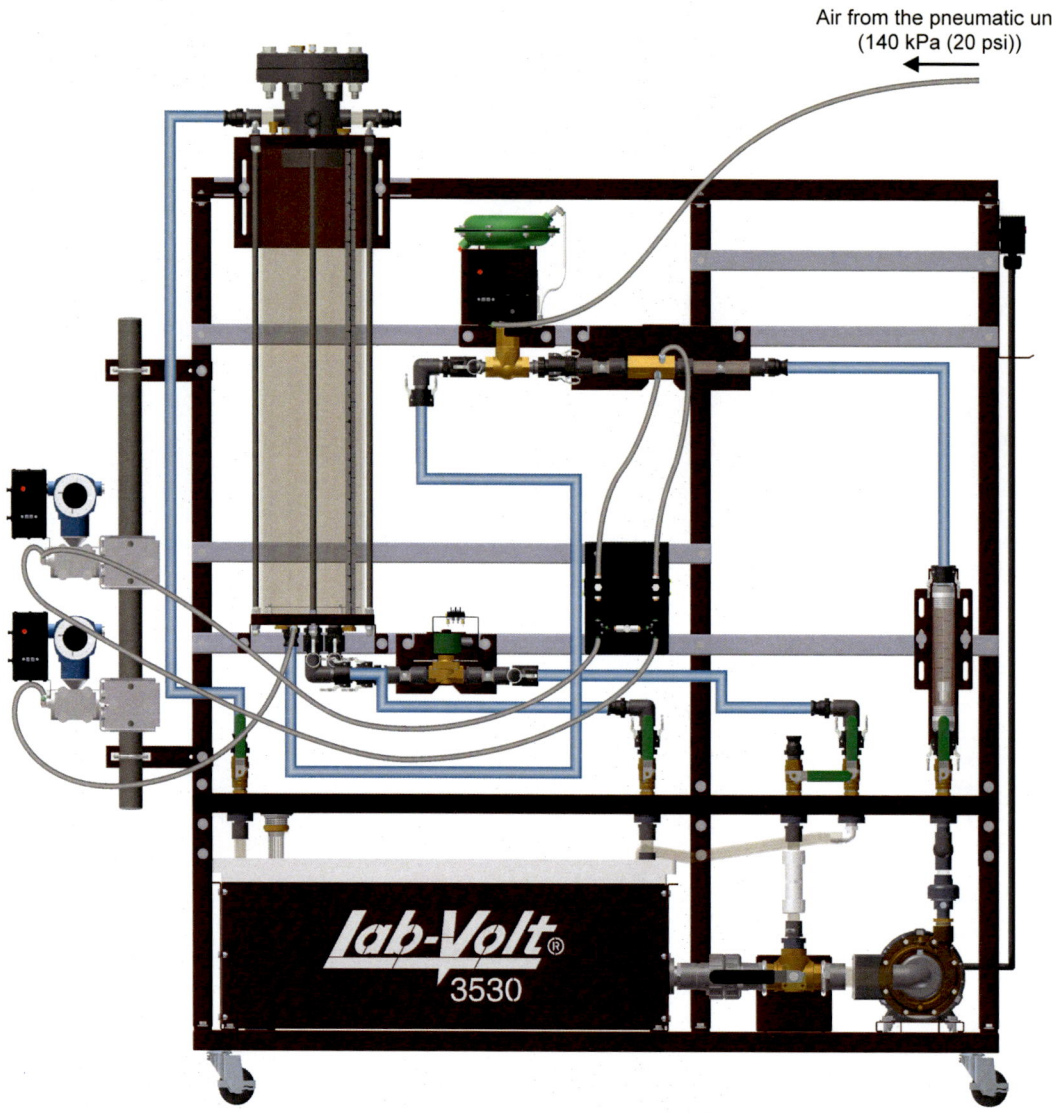

Figure 74. Setup – Cascade control loop.

2. Connect the control valve to the pneumatic unit.

3. Connect the pneumatic unit to a dry-air source with an output pressure of at least 700 kPa (100 psi).

4. Wire the emergency push-button so that you can cut power in case of emergency.

5. Do not power up the instrumentation workstation yet. You should not turn the electrical panel on before your instructor has validated your setup—that is not before step 11.

6. Connect the solenoid valve so that a voltage of 24 V dc actuates the solenoid when you turn the power on at step 11.

7. Connect the controller to the control valve and to the differential pressure transmitters. You must also include the recorder in your connections. On channel 1 of the recorder, plot the level output signal from the first transmitter. On channel 2, plot the flow output signal from the second transmitter. Channel 3 is used to display the output of the cascade control scheme sent to the control valve. Be sure to use the analog inputs of your controller to connect the differential pressure transmitters.

8. Figure 75 shows how to connect the different devices together.

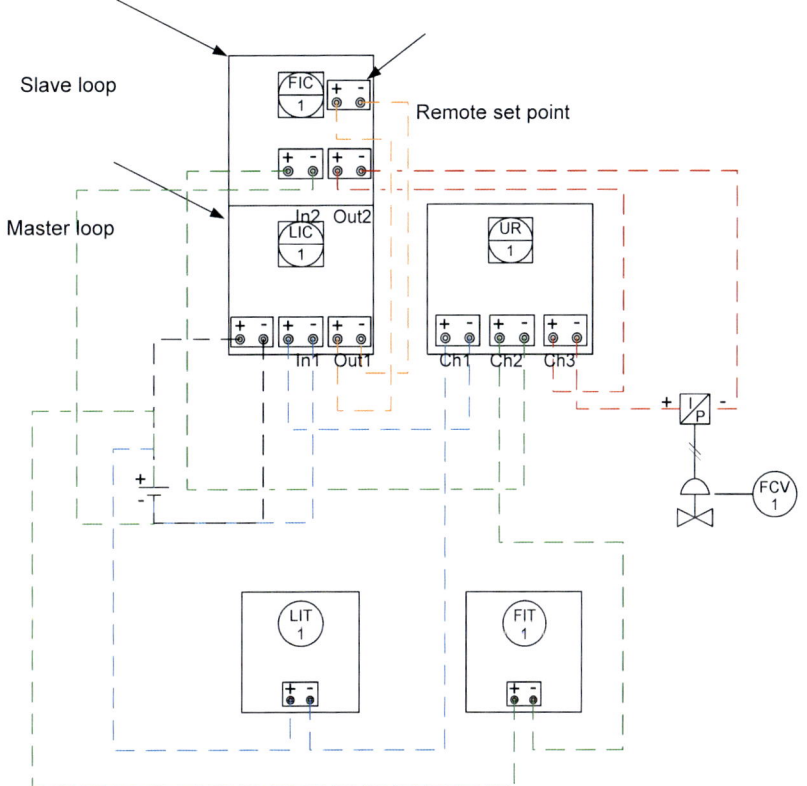

Figure 75. Connecting the instruments together for cascade control.

 The output of the master loop determines the set point of the slave loop in cascade control. This output is usually relayed internally to the slave loop when operating in cascade mode. However some controllers do require a physical connection from the output of the master loop to the remote set-point port of the slave loop. The corresponding electrical connections shown above (in orange) might or not be required with your controller (consult the technical information related to your controller).

9. Before proceeding further, complete the following checklist to make sure you have set up the system properly. The points on this checklist are crucial elements to the proper completion of this exercise. This checklist is not

> ▣ The solenoid valve under the column is wired so that the valve opens when the system is turned on.
>
> ▣ All unused ports on the column are capped.
>
> ▣ The ball valves are in the positions shown in the P&ID.
>
> ▣ The three-way valve at the suction side of the pump (HV1) is set so that the flow is directed toward the pump inlet.
>
> ▣ Valve HV3 is open to avoid pressurizing the column.
>
> ▣ The pneumatic connections are correct.
>
> ▣ The controller is properly connected to the differential pressure transmitters and to the control valve.

10. Ask your instructor to check and approve your setup.

11. Power up the electrical unit, this starts all electrical devices as well as the pneumatic unit. Activate the control valve of the pneumatic unit to power the devices requiring compressed air.

12. In manual mode, set the output of the slave controller to 0%, then 100%. The control valve should be fully open in the first case and fully closed in the second case. If it is not, revise the electrical and pneumatic connections and make sure the calibration of the I/P converter is appropriate.

13. Test your system for leaks. Use the drive to make the pump run at low speed to produce a small flow rate. Gradually increase the flow rate up to 50% of the maximum flow rate that the pumping unit can deliver (i.e., set the drive speed to 30 Hz). Stop the pump and repair any leak.

Adjusting the differential pressure transmitters

14. Bleed the impulse line and configure the **low-range** differential pressure transmitter for level measurement. Adjust the zero of the differential pressure transmitter.

    Set transmitter parameters so that a 4 mA signal is sent for a level of 0 m (0 in) and a 20 mA signal for a level of 0.75 m (30 in).

15. Connect the impulse lines of the **high-range** differential pressure transmitter to the three-valve manifold. Bleed the impulse lines and configure the transmitter for flow measurement. Adjust the zero of the differential pressure transmitter.

Set transmitter parameters so that a 4 mA signal is sent for a flow of 0 L/min (0 gal/min) and a 20 mA signal for a flow of 45 L/min (12 gal/min).

Tuning the slave loop

16. Tune the slave loop by using one of the methods presented in this unit (Trial-and-error method, Ultimate-cycle method, Open-loop Ziegler-Nichols method). Set the drive to **50 Hz**.

    Remember that the slave loop must be tuned independently of the master loop. You can do so directly with the appropriate loop on your controller or you can use a calibrator connected to the control valve if you want to characterize your process (disconnect the control valve from the controller while you characterize the process).

    Note the control parameters obtained for the slave loop below. It is recommended that the PI mode be used for a flow slave loop.

    $K_c =$

    $T_i =$

17. Set the slave loop according to the parameters you found in the previous step. Switch the slave control loop into Auto mode. This control loop should be kept in automatic mode for the remainder of this exercise.

    Test the flow control loop and fine tune the parameters if required.

Tuning the master loop

The master loop could be tuned with any of the three methods presented in this manual but two are recommended in the next step. The first method relies on the fact that a very similar process was tuned in the previous exercise and constitutes a good starting point. The second method is the open-loop Ziegler-Nichols method. Choose the one you prefer.

The action of the master loop will be inverted with respect to the slave loop as an increase of the measured level causes a decrease of the output variable, i.e. a decrease of the flow set point (slave loop).

18. **Trial-and-error method**: Program the master controller with the parameters of Ex. 2-3 and test it in cascade mode. Fine-tune the parameters using the trial-and-error method if required.

    Note the control parameters obtained for the master loop below.

    $K_c =$

    $T_i =$

If you prefer to characterize the level process and use the open-loop Ziegler-Nichols method, it is necessary to perform a step change in the output of the controller and to analyze the resultant data. Since the output of the master controller is directly linked to the set point of the slave controller, it is simpler to characterize the level process by performing a step change in the slave set point.

Open-loop Ziegler-Nichols method: Let the level process stabilize for a set point of the slave controller adjusted at 65%. Change the set point of the slave controller to 85% and wait for the level to stabilize again. (Be patient as this can take about twenty minutes). Transfer the data to a computer and determine the process parameters and the PID parameters:

$K_p =$ $\quad\quad\quad\quad t_d =$ $\quad\quad\quad\quad \tau =$

$\kappa = \left| \dfrac{\tau}{t_d K_p} \right| =$

Table 13. Ziegler-Nichols method – Master controller.

Mode	Proportional Gain K_c	Integral Time T_i	Derivative Time T_d
PI	$K_c = 0.9\,\kappa =$	$T_i = 3.33\, t_d =$	-
PID	$K_c = 1.2\,\kappa =$	$T_i = 2\, t_d =$	$T_d = 0.5\, t_d =$

Be careful with units. Your controller might use units different from those you used in Table 13.

Program the master controller with the parameters you just calculated and test it in cascade mode. Fine-tune the parameters if required.

Controlling the level/flow cascade loop

19. With your cascade controller properly set, create a 40% to 60% step change in the set point and observe the evolution of the system. Record and transfer the data to a computer. Plot a graph of your results.

20. Try different set-point values by increasing or decreasing the set point by step changes of 20%. Determine whether the controller tuning remains acceptable over a broad range of set points. Adjust the parameters if necessary.

21. Once the system is well tuned, set the controller to a set point of 50%.

    Create a sudden change in the process load by closing valve HV4. Is the controller able to rapidly correct for the load change without oscillation of the controlled variable?

    _____
    _____

    How much time is required for the process to return to the set point?

    _____

How does cascade control compare to conventional control with regard to the load change?

22. Re-open valve HV4 and let the process stabilize again to the set point of 50%. Suddenly increase the drive speed to 60 Hz to create a disturbance in the flow input.

    Is the system reacting in the same way as in the previous step? Can it handle the load change? How much time is required for the process to return to the set point?

23. Stop the pump and empty the column.

24. Stop the system, turn off the power, and store the equipment.

CONCLUSION

This exercise let you explore the possibilities and peculiarities of cascade control in the classical case of a level/flow process. You observed the amelioration of the overall response of the control scheme to a disturbance at the expense of a slightly higher complexity. This exercise also concludes the unit on basic techniques of process control. The next unit will put together the concepts you have learned so far by focusing on the troubleshooting of control loops.

REVIEW QUESTIONS

1. What determines the set point of the slave controller in a cascade control scheme?

2. What is the main purpose of cascade control?

3. What is the minimum requirement for the input (controlled) variables of the master and slave controllers?

Ex. 2-4 – Cascade Control of a Level/Flow Process ◆ *Review Questions*

4. In order for cascade control to be effective, should the slave process be more responsive or less responsive than the master process? Explain.

5. Briefly describe the procedure to tune a cascade control system.

Unit Test

1. What characterizes feedback control?

   a. It is a form of control where a change in the controlled variable triggers a corrective reaction of a manipulated variable.
   b. Controlled variables are not used in feedback control schemes.
   c. A disturbance must propagate through the process before the system can compensate for the deviation.
   d. Both a and c.

2. A well-tempered controller is operating in reverse action as the controlled variable keeps increasing above the set point. The manipulated variable will

   a. increase.
   b. decrease.
   c. stay the same.
   d. collapse.

3. What should the PI&D symbol be for the apparatus identified by a question mark, assuming it has no indicator?

   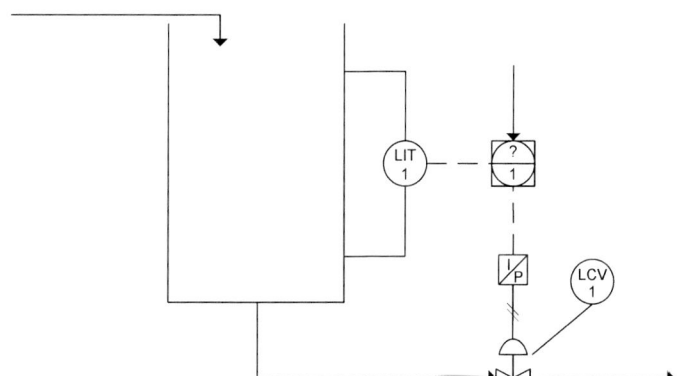

   a. PIC
   b. LC
   c. LIC
   d. JCVD

4. In on-off control, what is the name of the parameter used to reduce the oscillation frequency and what is the related phenomenon observed in the input-output relationship called?

   a. Dead band, Hysteresis
   b. Proportional band, Hysteresis
   c. Hysteresis, Noumenon
   d. Assertion, Sophism

5. What is the use of integral term?

   a. It increases the stability of the control scheme.
   b. It adds a predictive capability on the future state of the error.
   c. It eliminates the offset left by the proportional term.
   d. It makes the controller's equations more complex.

6. What should be the PI&D symbol be for the apparatus identified by a question mark?

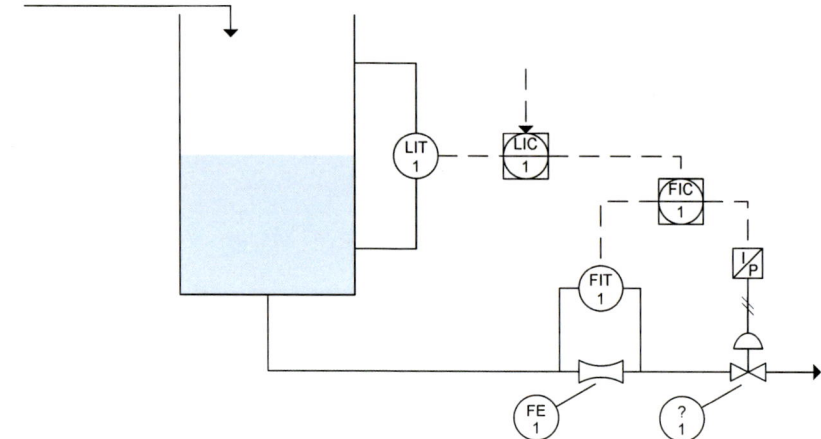

   a. LCV
   b. XKCD
   c. PCV
   d. FCV

7. The initial reaction of a PID controller following a sudden change in the set point is determined essentially by

   a. the proportional action.
   b. the integral action.
   c. the derivative action.
   d. Both a and c.

8. How is the ultimate gain K_u defined?

   a. K_u is the largest value of K_c (in P mode) such that the process is stable.
   b. K_u is the smallest value of K_c (in P mode) such that the process oscillates in a damped fashion.
   c. K_u is the largest value of K_c supported by your controller.
   d. K_u is the square root of the ultimate period.

9. A technician is using the open-loop Ziegler-Nichols method to tune a process ($K_p = 2$, $t_d = 5\ s$, and $\tau = 2\ min$) in PI mode. Somehow, he doesn't seem to carry the calculations correctly. Can you help him?

Open-loop Ziegler-Nichols method.

Mode	Proportional Gain K_c	Integral Time T_i	Derivative Time T_d
P	$K_c = \kappa =$	-	-
PI	$K_c = 0.9\ \kappa =$	$T_i = 3.33\ t_d =$	-
PID	$K_c = 1.2\ \kappa =$	$T_i = 2\ t_d =$	$T_d = 0.5\ t_d =$

where $\kappa = \left|\dfrac{\tau}{t_d K_p}\right|$

a. $K_c = 0.24$ $T_i = 10\ s$ $T_d = 2.5\ s$
b. $K_c = 14.4$ $T_i = 10\ s$ $T_d = 2.5\ s$
c. $K_c = 10.8$ $T_i = 16.65\ s$
d. $K_c = 0.18$ $T_i = 16.65\ s$

10. Cascade control

    a. increases the effect of disturbances on the secondary controlled variable.
    b. minimizes the disturbances that affect the secondary controlled variable before they cause pronounced changes in the primary controlled variable.
    c. requires the master process to be more responsive than the slave process to provide good control performance.
    d. requires the master controller to be tuned before the slave controller.

Unit 3

Troubleshooting a Process Control System

MANUAL OBJECTIVE

Give you an overview or the industrial troubleshooting process.

DISCUSSION OUTLINE

The Discussion of Fundamentals covers the following points:

- Troubleshooting
- Plant shutdown

DISCUSSION OF FUNDAMENTALS

Troubleshooting

Industrial equipment runs for prolonged periods of time. Such an extensive use has a toll on the equipment and, once in a while, one or more pieces of equipment break down. The main goal of **troubleshooting** is to track down failures that cause the plant or process loop to produce poor yield or force them to halt and correct the problem, returning the plant to its normal operation. Troubleshooting can also be used to improve process loops and reduce maintenance costs by preventing equipment failure.

Troubleshooting is a problem solving process. It is the process used to address and solve problems through logic and sound knowledge of the equipment and of the process. What makes process troubleshooting so difficult is that problems in a control loop can arise from different sources. There is no step-by-step procedure that provides the ability to find and correct all problems. Your best asset in troubleshooting is extensive knowledge of the control loop and its equipment.

There are general rules that apply to most troubleshooting situations. These rules should be viewed as guidelines. The table below gives the main steps that you should follow to efficiently troubleshoot a process control loop.

Step	Description	Example
Observe	This is the first step to troubleshooting. Determine what is wrong with the process compared to the normal operation of the system. Be sure to observe all the symptoms before trying to correct anything.	The controller input always reads 0%. Check the historical trend, the fault record, the pilot lamps, the LEDs, etc.
Analyze the available information	Analyze the available data and determine if you have enough information to form an hypotheses on the potential problem. Sometimes, more than one device may be defective. The defective part may also be located downstream or upstream of the instrument that behaves strangely.	The controller displays an error message, but it is the transmitter connected to the input that is defective.

Step	Description	Example
Acquire additional data	Collect all additional data available. Insufficient data may lead to the wrong hypotheses and ultimately to loss of time and money.	Check the status of the transmitter. Ask questions of the operator, he is a valuable source of information.
Identify potential problems and solutions	From the acquired data, form one or more hypotheses about the potential problems and solutions. Prioritize them.	Hypothesis/solution: • The controller input is defective/repair controller. • The transmitter does not work/change transmitter. • Someone entered the wrong configuration parameter on a device/restore configuration.
Test your hypotheses (trial and error)	Using trial and error, test your hypotheses starting with the one that is most likely.	Tests: • Check the controller input. • Replace the transmitter with a functional one. • Check the configuration parameters.
Observe the result	Observe the results of your test and revise your hypotheses, if required.	A reset on the parameters of the transmitter, may solve the problem.
Long range implementation	Once the problem(s) is (are) identified, plan for a long term corrective action that will reduce the risk of the same failure occuring again.	Implement a standard procedure to ensure the devices are always correctly configured.

The section below gives an example of the troubleshooting of a flow process.

Plant shutdown

Many plants run continuously to maximize production. Any interruption in the production costs a lot of money. Therefore, the plant managers try not to stop the production process except for an emergency or a scheduled maintenance shutdown, such as the annual shutdown.

It is during a plant shutdown that the troubleshooting skills of technicians and operators are challenged. The annual shutdown is the occasion to replace old equipment, make changes to programs and the configuration of devices, calibrate instruments, etc. Within 24 or 48 hours, workers try to do everything they cannot do while the plant is running. The crucial moment of a shutdown is when the workers try to startup the plant again. This is usually not a process that goes smoothly. When everything is started up, all sorts of problems are exposed and the time is limited for technicians to fix everything.

This section presents a troubleshooting scenario where a problem occurred in a flow control loop after a plant shutdown. It describes the steps that an experienced technician would follow to solve the problem. The scenario is divided into several subsections. Each subsection represents a step in the normal troubleshooting sequence. This scenario should give you an overview of the variety of elements that may cause a single problem in a control loop. In reality, the troubleshooting process may be far more complex than described below.

Description of the situation

After a plant shutdown, the operator screen indicates that the flow is low in the fourth control loop of the third sub-unit of the plant. This control loop is a flow control loop. Figure 76 shows the P&ID diagram of this control loop. As soon as the problem is detected, the plant technician responsible for this control loop begins troubleshooting the loop.

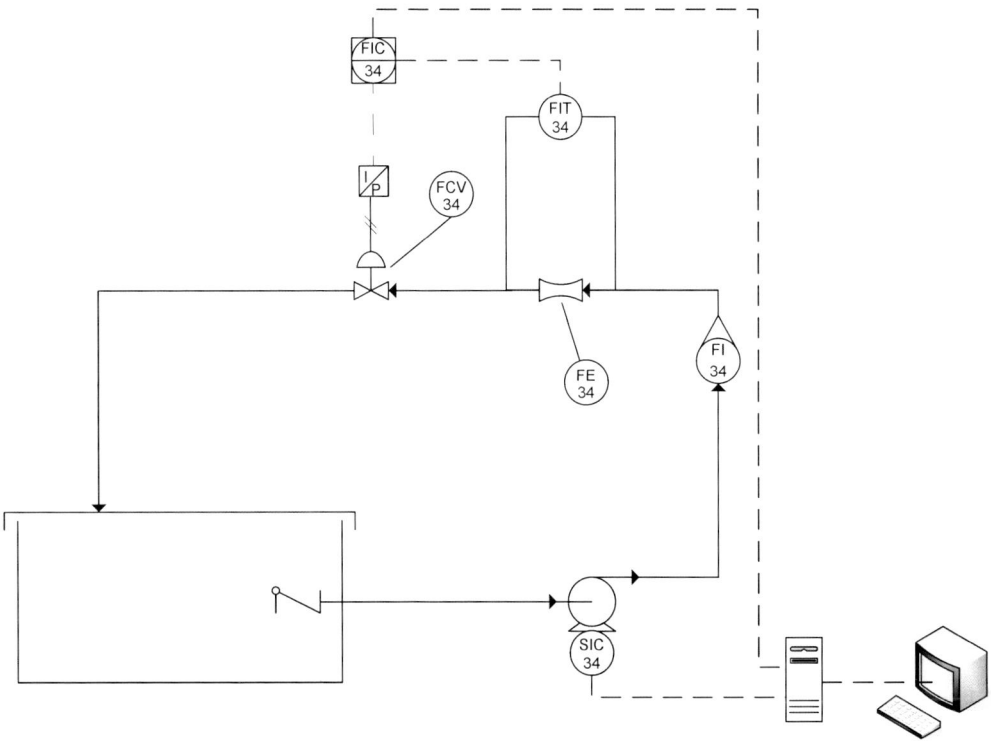

Figure 76. Flow control loop.

Observe

The first step for the technician is to observe the symptoms of the problem. He takes a look at the alarm on the operator screen. He takes note that the "Low flow alarm" indicates that the flow signal to the controller is below 10% of the span. He also checks when the alarm occurred, browses the historical trend, checks if the user interface displays a low flow or no flow at all. He gathers all the information available from the operators. The most important thing at this point is probably listening to the operators. In many situations, the operators have the key to the problem, listen to them, and take note of their observations.

Analyze the available information

From the information he has gathered, preliminary theories already form in the technician's mind. However, he knows that even if the alarm had been triggered by the controller, the problem may not come from this device. The problem may also originate from the pumping system, the transmitter, or the human machine interface (HMI).

Acquire additional data

The technician decides to collect additional data on the pumping system, transmitter, and HMI. He also decides to check if maintenance has been done on some of the devices of the flow control loop during the shutdown. If yes, he will check these devices first.

Identify potential problems and solutions

By collecting data, the technician may already have eliminated some possible causes. He narrows down his list of potential source(s) for the problem and gets ready to test his hypotheses.

Test your hypotheses (trial and error)

Once the technician has sufficient data he prioritizes the list of possible problems. He divides the possible problems into four categories: problems with the process (valves, pump, motor, etc.), problems with the instrumentation (transmitter), problems with the controller, and problems with the HMI. The potential problems that are the easiest and fastest to check are at the top of his list.

Table 14. Troubleshooting priority chart.

Category	Possible source	Things to check	Priority
Process	Valve(s)	• Defective valve(s) • Closed valves(s)	1
	Motor	• Motor power supply • Lockout device on the power supply • Motor tripped • Power supply cord • Motor interlock • Broken motor	2
	Pump	• Pump coupling • Rotation direction (phase inversion) • Impeller correctly installed	3
	Pump suction	• Strainer • Solids in the pump • Check valve	4
Instrumentation	Flow transmitter connections	• Transmitter power supply • Plugged impulse line • Disconnected impulse line • Transmitter wiring	5
	Flow transmitter calibration	• Transmitter calibration • Defective flow element	6

Category	Possible source	Things to check	Priority
Controller	Defective flow controller	• Controller power supply • Controller input • Controller wiring	7
	Controller configuration	• Configuration of the controller input • PID configuration	8
HMI	Human-machine interface	• OPC server configuration • Interface tags • Network connection between the server and the workstation • Workstation configuration • HMI program	9

Observe the result

The technician checks the various elements listed above one-by-one. If the system shows no improvement after testing one element, he checks the next on the list. Once he has identified the problem, he takes immediate action to correct it. Once a problem is fixed, the technician has to check if the loop responds correctly. Fixing a problem does not ensure that the loop will run correctly. Maybe there is more than one problem or defective device in the loop. If this is the case, fixing one problem may have changed how the whole system reacts, therefore, the technician may have to repeat the troubleshooting process, starting with the observation step.

Documenting

The technician keeps a record of the problem. Documenting the problem improves the global knowledge of the plant. It also gelps to identify devices that repeatedly fail and indicates where further investigation is required.

Long range implementation

Since a control loop that is down for several hours results in a net loss of money, plant managers want to prevent such a situation as much as possible. Long range solutions to avoid recurrent problems are essential.

As part of the long range implementation, the cause analysis is an important step for any plant manager. The cause(s) of a problem are usually one or more of the following:

- People
- Methods
- Material
- Equipment
- Environment

Once the cause(s) is (are) identified, it is important to decide what must be modified to prevent the problem from occuring again. A person or a group of persons must be in charge of the implementation of this modification within a fixed time frame. This modification, or improvement, may include establishing maintenance standards, improving employe training, instituting a monitoring program, and/or changing the equipment. Once the modification is implemented, the results of the implementation must be analyzed and, if they are satisfactory, the plant manager should consider standardizing this implementation on all the control loops to improve the overall performance of the plant.

Exercise 3-1

Guided Process Control Troubleshooting

EXERCISE OBJECTIVE   Develop your troubleshooting skills via a guided troubleshooting exercise.

DISCUSSION OUTLINE   The Discussion of this exercise covers the following point:

- Setting the scene

DISCUSSION

Setting the scene

This exercise casts you as the person in charge of a production unit in a chemical plant. This production unit includes several control loops. The main control loop of this unit is a flow control loop.

Today the plant is scheduled for a maintenance shutdown. Everything goes smoothly during the shutdown and you had plenty of time to do the maintenance of your production unit. Once the shutdown is over, you enthusiastically expect to instantly restart your control loop and resume normal operation of the plant. Unfortunately, nothing seems to be working the way it should and you scramble to fix the problem while the whole plant is paralyzed by your inactive process loop. Your boss comes by to inquire about the situation and he is very interested to know how much more time will be needed to resolve the problem...

Thankfully, you remain calm and you have a structured approach to track and correct the anomalies. By using the troubleshooting guidelines presented in this unit, you manage to find the problems and to get the plant back in operation rapidly.

PROCEDURE OUTLINE   The Procedure is divided into the following sections:

- Set up and connections
- Adjusting the differential pressure transmitter
- Adjusting the control flow loop
- Troubleshooting

PROCEDURE

Set up and connections

1. Connect the equipment according to the piping and instrumentation diagram (P&ID) shown in Figure 77 and use Figure 78 to position the equipment correctly on the frame of the training system.

Table 15. Material to add to the basic setup for this exercise.

Name	Model	Identification
Differential pressure transmitter (high-pressure range)	46920	PDIT 1
Controller	*	FIC
Flow control valve	46950-**	FCV
Venturi tube	46911	FE 1
Three-valve manifold	85813	-

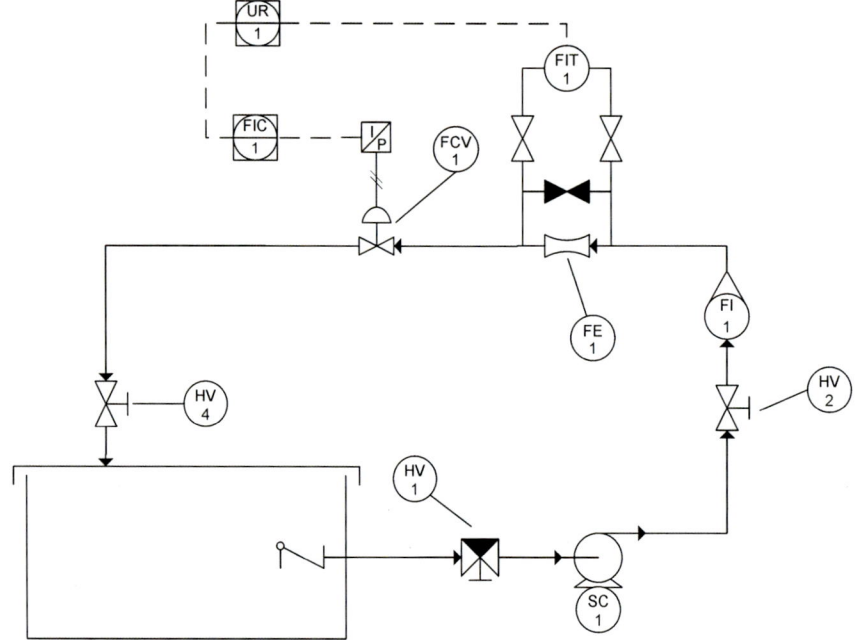

Figure 77. P&ID – Flow control loop.

Ex. 3-1 – Guided Process Control Troubleshooting ♦ *Procedure*

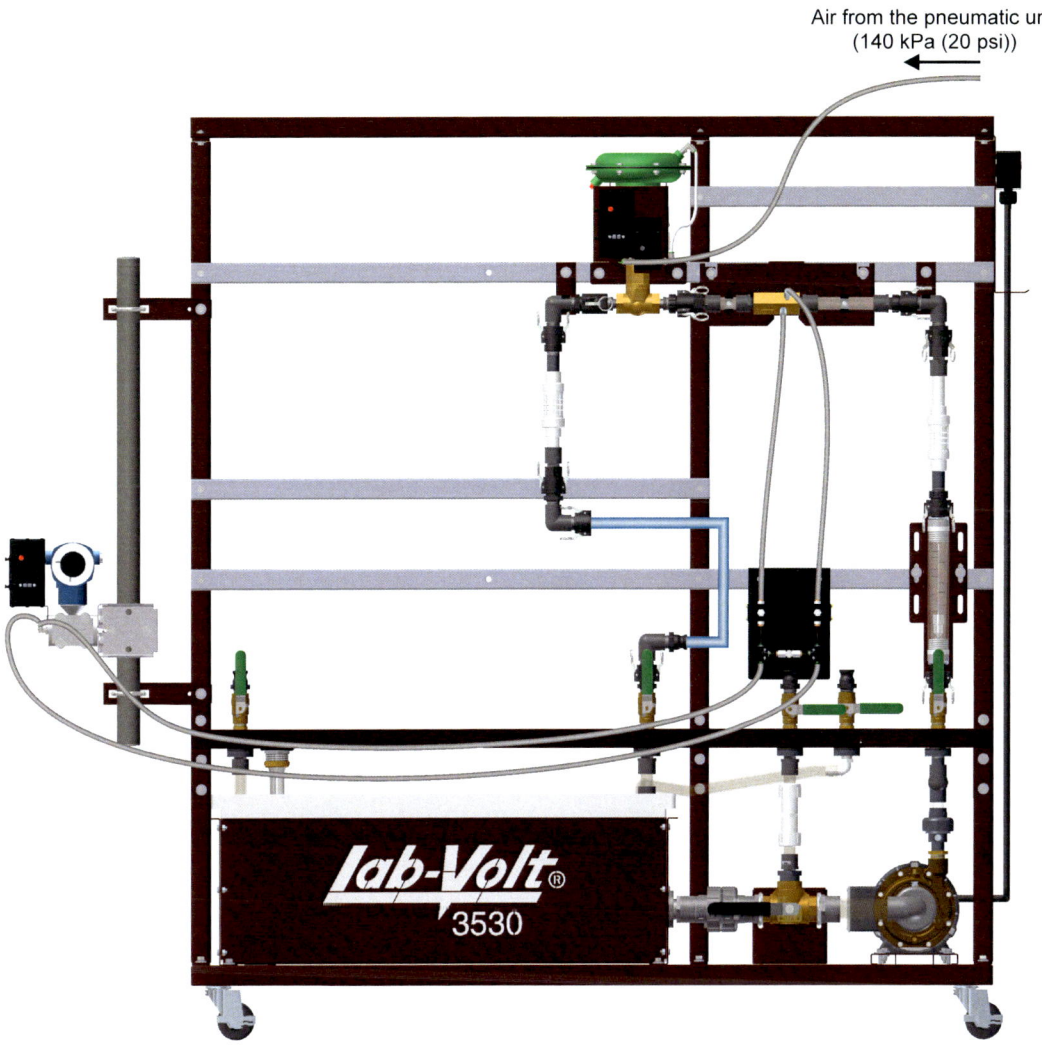

Figure 78. Setup – Flow control loop.

2. Connect the control valve to the pneumatic unit.

3. Connect the pneumatic unit to a dry-air source with an output pressure of at least 700 kPa (100 psi).

4. Wire the emergency push-button so that you can cut power in case of emergency.

5. Do not power up the instrumentation workstation yet. You should not turn the electrical panel on before your instructor has validated your setup—that is not before step 10.

6. Connect the controller to the control valve and to the differential pressure transmitter. You must also include the recorder in your connections. On channel 1 of the recorder, plot the output signal from the controller and on channel 2, plot the signal from the transmitter. Be sure to use the analog input of your controller to connect the differential pressure transmitter.

7. Figure 79 shows how to connect the different devices together.

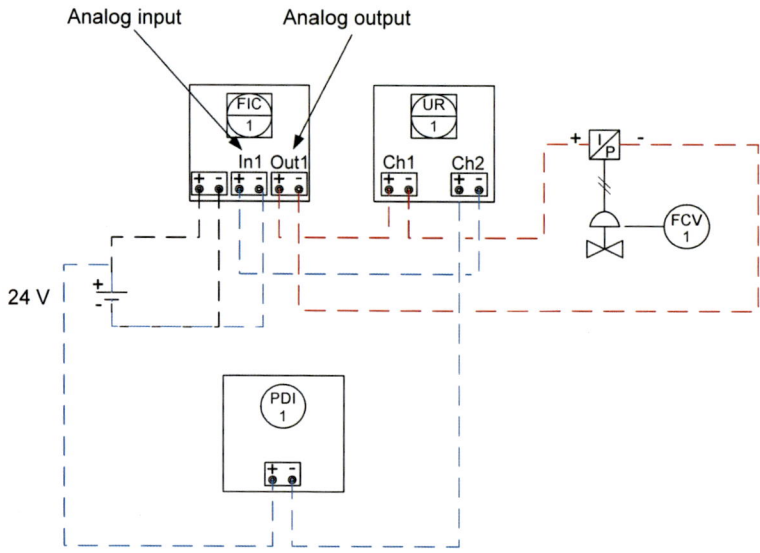

Figure 79. Connecting the equipment to the recorder.

8. Before proceeding further, complete the following checklist to make sure you have set up the system properly. The points on this checklist are crucial elements to the proper completion of this exercise. This checklist is not exhaustive, so be sure to follow the instructions in the *Familiarization with the Instrumentation and Process Control Training System* manual as well.

- The Venturi tube and the three-valve manifold are properly used according to the guidelines stated in the *Familiarization with the Instrumentation and Process Control Training System* manual.
- The ball valves are in the positions shown in the P&ID.
- The three-way valve at the suction side of the pump (HV1) is set so that the flow is directed toward the pump inlet.
- The control valve is fully open.
- The pneumatic connections are correct.
- The controller is properly connected to the differential pressure transmitter and to the control valve.
- The paperless recorder is connected correctly to plot the appropriate signals on channel 1 and channel 2.

9. Ask your instructor to check and approve your setup.

10. Power up the electrical unit, this starts all electrical devices as well as the pneumatic unit. Activate the control valve of the pneumatic unit to power the devices requiring compressed air. Test your system for leaks and for proper operation.

Adjusting the differential pressure transmitter

Be sure to use the differential pressure transmitter, Model 46920. This differential pressure transmitter has a high-pressure range.

11. Be sure to connect the impulse lines of the differential pressure transmitter to the three-valve manifold. Bleed the impulse lines and configure the transmitter for flow measurement. Adjust the zero of the differential pressure transmitter.

    Set transmitter parameters so that a 4 mA signal is sent for a flow of 0 L/min (0 gal/min) and a 20 mA signal for a flow of 40 L/min (10 gal/min).

Adjusting the control flow loop

12. With the drive set to 40.0 Hz, tune the controller in PI mode to regulate the flow. Use any of the three methods presented in this manual. Fine-tune the parameters and test your system at different set points.

    This is the baseline state of operation of your plant. Note the control parameters obtained for the master loop below.

    $K_c =$

    $T_i =$

13. Stop the system and leave the room while your instructor inserts a few faults in your control loop.

Troubleshooting

The objective of the upcoming manipulations is to identify the faults inserted into the system by your instructor. Follow the general guidelines and investigate any possible problems until you can pinpoint and explain the nature of the fault to your instructor. Try to get to the bottom of the problem and do not assume that the simplest explanation is necessarily the cause of the malfunction.

14. Slowly restart the process loop by starting the drive and progressively increasing its speed from 0 to 40.0 Hz. Make sure the controller is in the automatic mode.

Is the process behaving normally? What anomalies do you observe if any?

_____

_____

_____

15. Stop the drive for now. The steps presented in Table 14 are to be followed in order to localize the faults. The first elements to verify are the ones directly related to the process. As the motor and pump are not tampered with in this exercise, the valves are the only suspects in the Process category.

Inspect all hand-valves and make sure they are properly open or closed as indicated in the P&ID of Figure 77.

Test the control valve with a calibrator. The control valve should be fully open when it receives a signal of 20 mA and fully closed for a signal of 4 mA.

What is defective and why?

_____

_____

Confirm your diagnosis with your instructor who will deactivate the fault if you are right. Test the system to make sure everything related to the valves is now in working order.

16. The next step is to verify the different instruments connected to the process, in this case the differential pressure transmitter.

Start by verifying the power supply, the state of the impulse lines, and the wiring of the instruments.

The next point concerns the calibration of the apparatus. Make sure the flow reading is correct.

What is not working properly? Why?

_____

_____

Confirm your diagnosis with your instructor who will deactivate the fault if you are right. Test the system to make sure everything is now fine with the instrument.

Ex. 3-1 – Guided Process Control Troubleshooting ♦ *Conclusion*

17. As the system is still not functioning optimally, it is necessary to move to the next step of our troubleshooting sequence: the controller.

    At first glance, what would you say is wrong with the controller based on the behavior you have observed so far?

    _____

    Go through the list of points to verify until you can make the controller behave as well as before the insertion of faults in the system.

    What is not working properly? Why?

    _____

    _____

    Confirm your diagnosis with your instructor and deactivate the fault. Test the system to make sure the controller is now in good working order.

18. Stop the system, turn off the power, and store the equipment.

CONCLUSION

The troubleshooting of control processes is a crucial activity for anyone working in the control industry. This exercise introduced a structured approach for the quick and efficient resolution of problems. As proficiency comes with experience, the next experiment will let you develop your skills on a defective process without guidance.

Exercise 3-2

Non-Guided Process Control Troubleshooting

EXERCISE OBJECTIVE Allow you to experience a troubleshooting situation without any guidelines.

DISCUSSION OUTLINE The Discussion of this exercise covers the following point:

- Non-guided troubleshooting

DISCUSSION **Non-guided troubleshooting**

In this exercise you take one of the setups from Unit 2 and troubleshoot it without further indication or help from the humble writers of this manual. The difficulty of this exercise depends of the fault(s) that your instructor inserts in the control loop. Ask for assistance or for further instructions from your instructor if required.

Keeping a structured approach and following the troubleshooting guidelines should help to troubleshoot your process loop.

PROCEDURE OUTLINE The Procedure is divided into the following sections:

- Set up and connections
- Troubleshooting

PROCEDURE **Set up and connections**

1. Your instructor will ask you to make the setup from one of the exercises of Unit 2. Follow all the instructions and the piping and instrumentation diagram (P&ID) of the corresponding exercise.

2. Do not forget to ask your instructor to check and approve your setup before turning the power on.

3. Power up the electrical unit and configure the transmitter(s) and the controller as specified in the exercise of Unit 2. Tune your controller and make sure your control loop is operational.

4. Stop the system and leave the room while your instructor inserts one or more faults in your control loop.

Troubleshooting

5. Use the troubleshooting sequence presented earlier in this unit to troubleshoot your control loop. Once you have identified a fault, report to your instructor. He will remove the fault and allow you to resume your troubleshooting sequence if any fault remains.

CONCLUSION

After this exercise, you may not be a troubleshooting expert, but you now know a trick or two.

Appendix A

I.S.A. Standard and Instrument Symbols

Introduction

Instrumentation systems can be incredibly complex. Without the use of a standardized method of symbols, technicians and engineers would be at a loss when confronted with a new system. The *Instrumentation Symbols and Identification* standard, prepared by the International Society of Automation (ISA), is the primary method that has been adopted throughout the process industry. This standard must be like a second language to anyone working in the field of instrumentation. Throughout this document, and all the teaching material of the Instrumentation and Process Control System, the ISA standard is used in the examples and in the process flow diagrams. This section is a brief introduction to the basic set of rules and symbols of the ISA *Instrumentation Symbols and Identification* standard.

A process flow diagram is a representation of the flow of processes and equipment. A flow diagram usually includes:

- General instrument symbols (or function symbols)
- Function designation symbols
- Tag numbers
- Instrument line symbols
- Specific component symbols

Figure 1 shows a typical process flow diagram that includes these five elements.

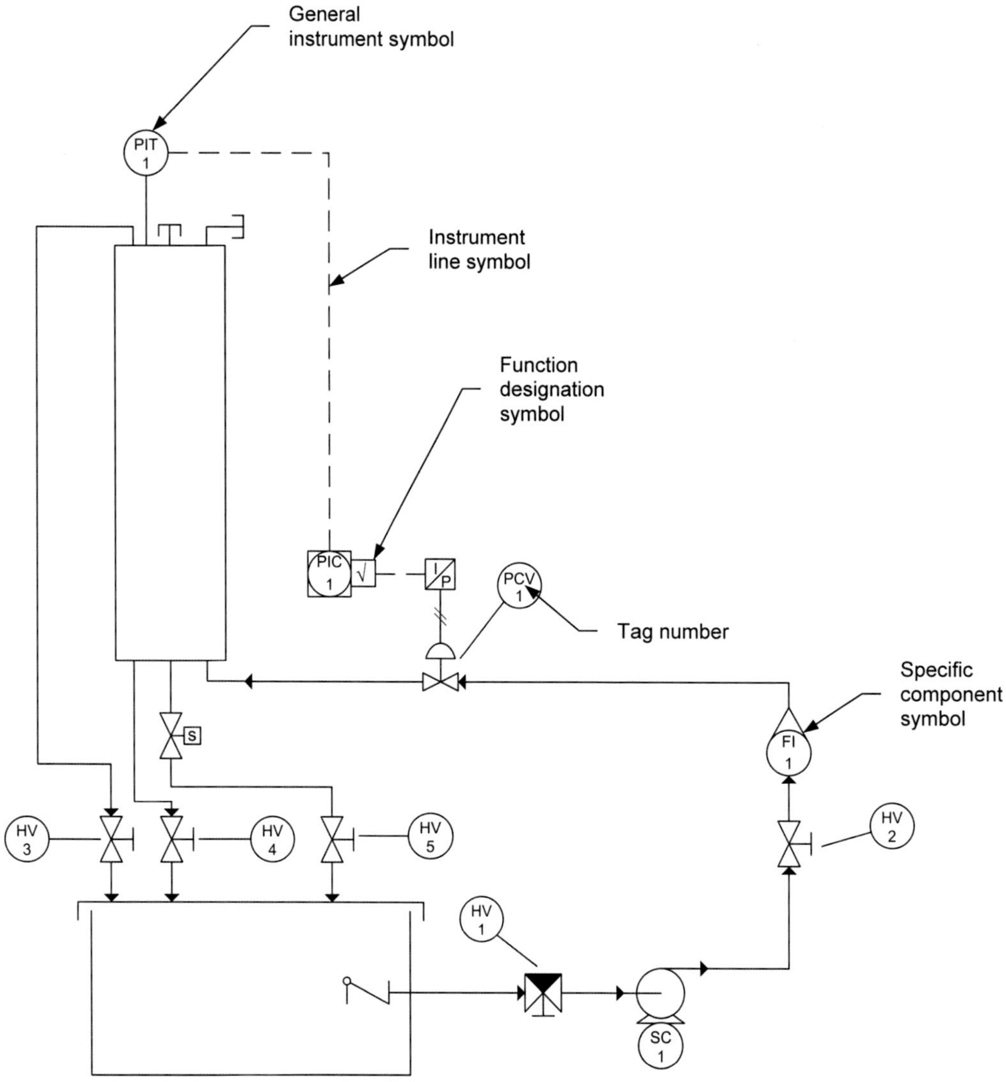

Appendix A

I.S.A. Standard and Instrument Symbols

Figure 1. Typical process flow diagram.

Tag numbers

In a flow diagram, an alphanumeric code, called a tag number, identifies the different instruments. A tag number consists of a functional identification and of a loop number.

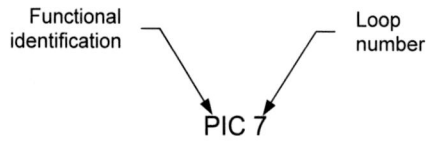

Appendix A

I.S.A. Standard and Instrument Symbols

The functional identification, as its name indicates, gives the function of the instrument and is not related to the construction of the instrument. The first letter of the functional identification designates the measured variable or initiating variable. The succeeding letter(s) designate the function of the instrument. For example, FT can designate a differential-pressure transmitter that measures a flow, LT can designate a differential-pressure transmitter that measures a level, and PDT can designate a differential-pressure transmitter that measures a pressure. A modifying letter can be added after the first letter or after the succeeding letter(s) to modify its meaning. The letters in a functional identification are all uppercase and their number should not exceed four.

In a typical tag number, a loop number usually follows the functional identification. The loop number indicates the loop to which the component belongs. Large systems are divided into sub-units; in that case, the first two digits of the loop number represent the identification number of the sub-unit and the last digit identifies the control loop within the sub-unit. For example, 112 may represent the second control loop of sub-unit 11. When more than one instrument has the same functional identification within a loop, a suffix may follow the identification number (e.g., PI-1A and PI-1B) or the instruments' identification numbers may be sequential (e.g., PI-1 and PI-2).

Table 1 gives the significance of the letters in the functional identification of a device. Table 2 follows with examples of how the letters and numbers are selected to form a complete tag number.

Table 1. Identification letters.

	First-letter		Succeeding-letters		
	Measured or Initiating Variable	Modifier	Readout or Passive function	Output function	Modifier
A	Analysis		Alarm		
B	Burner, Combustion		User's Choice	User's Choice	User's Choice
C	User's Choice			Control	
D	User's Choice	Differential			
E	Voltage		Sensor (Primary Element)		
F	Flow Rate	Ratio (Fraction)			
G	User's Choice		Glass Viewing Device		
H	Hand				High
I	Current (Electrical)		Indicate		
J	Power	Scan			
K	Time, Time Schedule	Time Rate of Change		Control Station	
L	Level		Light		Low

Appendix A — I.S.A. Standard and Instrument Symbols

	First-letter		Succeeding-letters		
	Measured or Initiating Variable	Modifier	Readout or Passive function	Output function	Modifier
M	User's Choice	Momentary			Middle, Intermediate
N	User's Choice		User's Choice	User's Choice	User's Choice
O	User's Choice		Orifice, Restriction		
P	Pressure, Vacuum		Point (Test) Connection		
Q	Quantity	Integrate, Totalize			
R	Radiation		Record		
S	Speed, Frequency	Safety		Switch	
T	Temperature			Transmit	
U	Multivariable		Multifunction	Multifunction	Multifunction
V	Vibration, Mechanical Analysis			Valve, Damper, Louver	
W	Weight, Force		Well		
X	Unclassified	X Axis	Unclassified	Unclassified	Unclassified
Y	Event, State, or Presence	Y Axis		Relay, Compute, Convert	
Z	Position, Dimension	Z Axis		Driver, Actuator, Unclassified Final Control Element	

Appendix A — I.S.A. Standard and Instrument Symbols

Table 2. Tag number examples.

ISA Tag Number	Identification
FE1	Primary Flow sensing Element, flow control loop 1
FT1	Flow Transmitter, flow control loop 1
FIC1	Flow Indicating Controller, flow control loop 1
SC1	Speed drive Control, control loop 1
PT3	Pressure Transmitter, pressure control loop 3
PIC3	Pressure Indicating Controller, pressure control loop 3
PY3	Current to Pressure converter (Y), pressure control loop 3
PCV3	Pressure Control Valve, pressure control loop 3
AE6, pH	Primary Analysis (pH) Element, pH control loop 6
AT6	pH Analysis Transmitter, pH control loop 6
ATC6	pH Analysis and Indicating Controller, pH control loop 6

Function designation symbols

Table 3 lists the function designation symbols. The *Instrumentation Symbols and Identification* standard recommends using these symbols as flags in detailed piping and instrumentation diagrams to designate the function of apparatuses such as controllers, computing devices, relays, or converters. These symbols are also used as function blocks in conceptual diagrams or in diagrams mixing both function blocks and instrumentation symbols.

Table 3. Function designation symbols.

Symbol	Function
Σ	Sum
Σ/n	Average
Δ	Difference
k, 1:1, 2:1	Proportional
$\int$	Integral
d/dt	Derivative

Appendix A — I.S.A. Standard and Instrument Symbols

Symbol	Function
×	Multiply
÷	Divide
$\sqrt[n]{\,}$	Root extraction
x^n	Exponential
f(x)	Nonlinear or unspecified function
f(t)	Time function
>	High selection
<	Low selection
⇥	High limit
⇤	Low limit
-k	Reverse proportional
⩔	Velocity limiter
+ - ±	Bias
/	Convert (from * to *, e.g., from in. to cm)
**H **L **HL	Signal monitor (The ** are to be replaced with the appropriate letters from Table 1)

Appendix A I.S.A. Standard and Instrument Symbols

General instrument symbols

General instrument symbols represent primary sensing elements, transmitters, indicators, gauges, and controllers. Table 4 lists the four types of instrument symbols that the *Instrumentation Symbols and Identification* standard recognizes. In the table below, a computer function is to be understood in a broader sense than a desktop computer. It is a device that performs calculations or logic operations and transmits the result as outputs. A shared display is a device which can display process information from a number of sources and a shared controller controls different variables at the same time.

Table 4. General Instrument symbols.

	Field Mounted	Main Control Panel	Auxiliary Location
Discrete Instrument	○	⊖	⊖
Computer Function	⬡	⬡	⬡
Shared Display/Control	◻	◻	◻
Programmable Logic Control	◇	◇	◇

Instrument line symbols

The line symbols are used to represent various connection lines between process components.

Appendix A — I.S.A. Standard and Instrument Symbols

Table 5. Line symbols.

Description	Symbol
Instrument Supply OR Connection to Process	———————
Undefined Signal	—/—/—
Pneumatic Signal	—//—//—
Electric Signal	– – – – – – – or —///—///—
Hydraulic Signal	—L—L—
Capillary Tube	—×—×—
Electromagnetic or Sonic Signal (Guided)	—∿—∿—
Electromagnetic or Sonic Signal (Not Guided)	∿ ∿
Internal System Link (Software/Datalink)	—o—o—
Mechanical Link	—•—•—
Pneumatic Binary Signal	—⁂—⁂—
Electric Binary Signal	– /– /– /– or —⁂—⁂—

Other component symbols

Other component symbols are used to identify specific primary sensing elements and final control elements, as Table 6 shows.

Table 6. Other Symbols.

Symbol	Description
(LE/Cap.) – ▶ (LIT) – ▶	Capacitive level transmitter
→⊘	Centrifugal pump
—⋈▶	Check valve

Appendix A — I.S.A. Standard and Instrument Symbols

Symbol	Description
	Conductivity level switch
	Conductivity transmitter
	Coriolis flow transmitter
	Differential-pressure transmitter
	Electric control valve
	Energy manager
	Float level switch
	Hand-operated valve (fully closed)
	Hand-operated valve (fully open)
	Hand-operated valve (partially closed)
	Heat exchanger
	Magnetic flow transmitter
	Metering pump
	Orifice plate

Symbol	Description
	pH transmitter
	Pitot tube
	Plate heat exchanger
	Pneumatic control valve
	Pneumatic control valve with a positioner
	Pressure gauge
	Pressure switch
	Radar level transmitter (guided)
	Radar level transmitter (non-guided)
	Solenoid valve
	Temperature transmitter (where TE is an RTD or a thermocouple)
	Three-way hand-operated valve (cross-flow position)

Appendix A

I.S.A. Standard and Instrument Symbols

Symbol	Description
	Three-way hand-operated valve (direct-flow position)
	Three-way pneumatic control valve
	Trend recorder/paperless recorder
	Turbine/Paddle-wheel flow transmitter
	Ultrasonic flow transmitter
	Ultrasonic level transmitter
	Variable-area flowmeter
	Venturi tube
	Vibrating level switch
	Vortex flow transmitter

Index

bias **40**, 42, 44, 45, 46, 65
block diagram **3**, 41, 50

capacitance 5, **6**, 7, 8, 9, 10, 12, 13, 17, 18, 33, 35
cascade control **87**, 88, 89
closed loop 1, **2**, 15, 54, 65, 66, 67, 78
control lag **11**
controlled variable .. **2**, 4, 31, 32, 33, 34, 35, 36, 37, 40, 46, 47, 53, 54, 56, 65, 68, 77, 78, 88, 89
controller gain **38**, 39, 40, 42, 44, 46, 48, 51, 53, 54, 56

dead band **36**, 37, 53
dead time **11**, 12, 15, 18, 19, 20, 21, 37, 77
derivative 8, 12, 32, 38, **39**, 40, 45, 46, 47, 49, 50, 54, 56, 65, 66, 89
derivative time **39**, 40, 46, 56, 65
differential equation **8**, 9, 10
direct action **32**, 36
disturbance **2**, 3, 31
dynamical system **1**, 5

error 20, 31, **32**, 38, 39, 40, 42, 43, 45, 46, 47, 48, 49, 53, 54, 57, 65, 68

feedback **31**
final control element **3**, 16, 53, 89
first-order response **8**, 10, 12, 19

graphical method **19**

hysteresis **36**

ideal controller **49**
inertia **6**
integral 32, 38, **39**, 40, 42, 43, 44, 45, 46, 47, 49, 50, 53, 54, 56, 65, 66, 89
integral time **39**, 40, 42, 43, 44, 46, 53, 56, 65
interacting controller **50**, 51

manipulated variable **2**, 4, 10, 17, 31, 32, 35, 39, 53
manual reset **65**
master controller **87**, 88, 89
master loop **87**, 89

noise **46**, 50, 89
non-interacting controller **49**, 50, 51
non-self-regulating **7**

on-off controllers **33**
open loop 1, **2**, 15, 16, 19, 68, 77, 78
oscillatory response **31**, 40, 56
overdamped 54, **68**

parallel controller **51**
PID controller **38**, 39, 45, 47
pre-act **45**

Index

primary element **3**
process gain **13**, 15, 18, 38, 77
process lag **11**
proportional ... 32, 38, **39**, 40, 41, 42, 45, 46, 47, 48, 50, 51, 53, 54, 65, 66, 67, 89
proportional band **38**, 42, 44, 46, 53, 54, 66

quarter-amplitude decay 65, **67**, 68, 78

rate action **45**
RC circuit **7**, 9, 10
reset rate **44**, 46
resistance **5**, 6, 7, 10, 12
resistive part **5**
response curve **7**, 8, 9, 10, 12, 13, 15, 16, 17, 18, 19, 46, 77
response time **13**, 56
reverse action **32**, 33, 35, 36, 37

scan time **45**
secondary element **3**, 16
self-regulating **7**, 15
serial controller **50**
set point **2**, 31, 32, 34, 35, 36, 37, 39, 40, 42, 46, 47, 48, 53, 54, 56, 65, 67, 68, 77, 78, 88
single-capacitance process 7, 8, 9, 10, 12, 13, 17, 18
slave controller **87**, 88, 89
slave loop **87**, 89
steady state 7, **13**, 17, 19, 40, 42, 47
step change **6**, 10, 11, 12, 13, 15, 16, 17, 18, 19, 47, 53, 54, 56, 65, 67, 68, 77

time constant 8, 10, **12**, 13, 15, 18, 19, 20, 21, 39, 42, 43, 44, 46, 77, 89
trial and error method **53**, 54
troubleshooting 3, 4, **103**, 104, 105, 107, 109, 117

ultimate method **65**
ultimate period **66**, 67
ultimate-cycle **65**, 66, 68
underdamped 56, **68**

Bibliography

BIRD, R. Byron, STEWART, W.E, and LIGHTFOOT, E.N. *Transport Phenomena*, New York: John Wiley & Sons, 1960
ISBN 0-471-07392-X

CHAU, P. C. *Process Control: A First Course with MATLAB*, Cambridge University Press, 2002.
ISBN 0-521-00255-9

COUGHANOWR, D.R. *Process Systems Analysis and Control*, Second Edition, New York: McGraw-Hill Inc., 1991.
ISBN 0-07-013212-7

LIPTAK, B.G. *Instrument Engineers' Handbook: Process Control*, Third Edition, Pennsylvania: Chilton Book Company, 1995.
ISBN 0-8019-8542-1

LIPTAK, B.G. *Instrument Engineers' Handbook: Process Measurement and Analysis*, Third Edition, Pennsylvania: Chilton Book Company, 1995.
ISBN 0-8019-8197-2

LUYBEN, M. L., and LUYBEN, W. L. *Essentials of Process Control,* McGraw-Hill Inc., 1997.
ISBN 0-07-039172-6

LUYBEN, W.L. *Process Modeling, Simulation and Control for Chemical Engineers*, Second Edition, New York: McGraw-Hill Inc., 1990.
ISBN 0-07-100793-8

MCMILLAN, G.K. and CAMERON, R.A. *Advanced pH Measurement and Control*, Third Edition, NC: ISA, 2005.
ISBN 0-07-100793-8

MCMILLAN, G. K. *Good Tuning: A Pocket Guide*, ISA - The Instrumentation, Systems, and Automation Society, 2000.
ISBN 1-55617-726-7

MCMILLAN, G. K. *Process/Industrial Instruments and Controls Handbook*, Fifth Edition, New York: McGraw-Hill Inc., 1999.
ISBN 0-07-012582-1

PERRY, R.H. and GREEN, D. *Perry's Chemical Engineers' Handbook*, Sixth Edition, New York: McGraw-Hill Inc., 1984.
ISBN 0-07-049479-7

RAMAN, R. *Chemical Process Computation*, New-York: Elsevier applied science ltd, 1985.
ISBN 0-85334-341-1

RANADE, V. V. *Computational Flow Modeling for Chemical Reactor Engineering*, California: Academic Press, 2002.
ISBN 0-12-576960-1

SHINSKEY, G.F. *Process Control Systems*, Third Edition, New York: McGraw-Hill Inc., 1988.

Bibliography

SMITH, Carlos A. *Automated Continuous Process Control*, New York: John Wiley & Sons, Inc., 2002.
ISBN 0-471-21578-3

SOARES, C. *Process Engineering Equipment Handbook*, McGraw-Hill Inc., 2002.
ISBN 0-07-059614-X

WEAST, R.C. *CRC Handbook of Chemistry and Physics*, 1st Student Edition, Florida: CRC Press, 1988.
ISBN 0-4893-0740-6